TRANSACTIONS

of the

American Philosophical Society

Held at Philadelphia for Promoting Useful Knowledge

VOLUME 75, Part 5, 1985

The Peccary—
With Observations on the Introduction of Pigs to the New World

R. A. DONKIN

Fellow of Jesus College Cambridge

THE AMERICAN PHILOSOPHICAL SOCIETY

Independence Square, Philadelphia

1985

Library of Congress Catalog
Card Number 84-45906
International Standard Book Number 0–87169–755–6
US ISSN 0065–9746

CONTENTS

MAPS

FIGURES

PREFACE

The three living species of peccary inhabit a vast area of the New World, between roughly 35 degrees of latitude north and south of the equator. They are primarily forest or woodland animals, but two species (one of them only recently discovered) have adapted to scrub-dominated ecosystems, both natural and anthropogenic, particularly around the latitudinal and altitudinal margins of their range. The overall distribution has contracted since the beginning of European settlement, as a result of intensive hunting and reduction in the extent of preferred habitats. Local depletion of the population also occurs for the same reasons. Nevertheless, peccaries are remarkably resilient animals; their comparatively low fertility rates are matched by low (natural) mortality, and all three species are unspecialized feeders. Cultivated plants add to a wide variety of natural foodstuffs. Like pigs, peccaries are good pioneers, and when hunting pressures are relaxed, for whatever reasons, numbers usually recover.

In traditional societies, the peccary is hunted chiefly for meat, and within the combined distribution of the species probably no other animal has contributed more to human food supply. Europeans have valued both the meat and, on a much larger scale, the hides. As far as is known, the peccary has never been domesticated, that is, bred regularly in captivity, but juveniles are often reared (for food), and some are tamed and treated as pets. Perhaps the conditions and processes that would have led to domestication were disrupted by the European conquest (and the introduction of the pig). At the same time, other kinds of relationship with animals may have partly substituted for domestication. These controversial questions are taken up in the concluding sections of the monograph.

The accompanying maps, mainly on a continental scale, serve to locate evidence referred to in the text. They represent states of knowledge, the pattern of discovery over time, and the adventitious recording and survival of information. New evidence (not to mention what has inadvertently been missed) will undoubtedly modify whatever tentative conclusions it has been possible to draw from the distributional (and other) record.

PART I: DISTRIBUTION, HABITAT, AND BIOLOGY

A: DESCRIPTION

The first scientific descriptions of the collared and the white-lipped peccary (*Tayassu tajacu* and *Dicotyles pecari*) were supplied by Félix de Azara, as part of a natural history of the quadrupeds of Paraguay, in 1801.[1] All previous accounts are to a varying extent unsatisfactory. The earliest known reference to "wild pigs of *two* kinds" is in Hans Staden's observations (1547–1555) on eastern Brazil.[2] However, even at this early date, one of the two may have been a feral European pig. Feral pigs were a source of confusion[3] until the early nineteenth century and sometimes gave rise to the opinion that there were three species or genera (*Catagonus* sp. of the Chaco was only identified in the mid-1970s and cannot be recognized in earlier, general descriptions of the New World pigs). Other early authorities, including Bernardino de Sahagún[4] and Francisco Hernández (ca. 1570),[5] described only one animal (Fig. 1), either one or other species, or combined some of the characteristics of each. Although strong evidence that there were at least two kinds of peccary was available in the published work of those with first-hand knowledge of the American tropics, such as Pierre Barrère (1741),[6] none of the leading systematists of the seventeenth and eighteenth centuries (John Ray,[7] M. J. Brisson,[8] Carl Linnaeus,[9] and the Comte de Buffon[10]) clearly distinguished between them.[11]

---

[1] Azara, 1801: 1: pp. 18–42, 1802: 1: pp. 23–29, 1838: 1: pp. 115–126. Azara visited South America in 1783.

[2] Staden (first published at Marburg, 1557), 1874: p. 160. Cf. Soares de Souza (1587), 1945: 2: p. 136; León Pinelo (ca. 1650), 1943: 2: pp. 50–51; Ruiz Blanco (ca. 1690), 1965: p. 23.

[3] Acuña (1639/1641), 1698: p. 69, 1859: p. 22; Ponce (1584–1592), 1875: 1: p. 90; Acosta (1570–1587), 1880: 1: pp. 273–274, 282–283, 1962: pp. 199–200, 205–206; Keymis (1596), 1904: p. 458.

[4] Sahagún, 1963: p. 10.

[5] Hernández, 1959: 1: pp. 310–311.

[6] Barrère, 1741: pp. 160–162 (French Guiana or Cayenne). J. G. Stedman (*Surinam*, 1796: 1: pp. 355–356) described two kinds of "wild swine" (apparently the collared peccary and the white-lipped peccary) and a third kind, the *cras-pingo*, which he believed, probably correctly, to have originated in Europe (feral *Sus*) or Africa. Antonio Caulín ([1779], 1980: 1: p. 73) also referred to "tres especies de puercos monteses" in Nueva Andalucía (Venezuela).

[7] Ray, 1693: p. 97.

[8] Brisson, 1756: p. 111.

[9] Linnaeus, 1758–1759: 1: p. 50, 1767–1770: 1: p. 103, [J. F. Gmelin], 1788–1793: 1: p. 219. Cf. Kerr [Linnaeus/Gmelin], 1792: pp. 352–353; Link, 1794–1801: 2: pp. 104–105.

[10] Buffon (ca. 1780), 1884: 9: pp. 233–235, 10: pp. 404–406.

[11] Buffon (ibid.) came closest, but was in doubt about the status of the larger animal reported by De La Borde from Cayenne. On Buffon, see Azara (1801), 1838: 1: p. 122.

COYÁMETL O QUAUHCOYÁMETL

Fig. 1. The peccary (? collared). Francisco Hernández (ca. 1570), 1959: 2: 310–311.

Peccaries superficially resemble the Old World pigs, to which of course they are related (superfamily Suiformes). A stocky body, massive head, and short, thin legs are their more obvious physical characteristics. The white-lipped peccary is the larger of the two, adult specimens (of both sexes) weighing 50 to 75 pounds or more.[12] White cheeks, snout, and lips contrast sharply with the gray-black color of the rest of the body. The more numerous races of collared peccary probably vary more in weight (average about 40 pounds).[13] Their distinctive characteristic is a narrow, semicircular "collar" or diagonal band of lightish hairs, set against a pelage of medium gray (black mixed with white or tawny). The earliest known reference to the "collar blanco" is in Francisco Montero de Miranda's *Descripción de la provincia de la Verapaz* (1574).[14] Juvenile peccaries are reddish brown in color with less distinctive markings.[15]

Both species, contrary to the opinion of some early authors,[16] have a dorsal gland about seven centimeters in diameter, first described as a "navel" (*ombligo*).[17] This emits a light brown, musky substance, especially

---

[12] Gilmore (1950), 1963: p. 382; Nietschmann, 1973: p. 165 (73 pounds). Cf. Enders, 1935: p. 477 (perhaps up to 100 pounds).

[13] Gilmore (1950), 1963: p. 382 (30 to 45 pounds); Crandall, 1964: p. 528 (maximum 65 pounds); Leopold (1959), 1972: p. 493 (30 to 55 pounds); Nietschmann, 1973: p. 165 (55 pounds).

[14] Montero de Miranda, 1953: p. 348.

[15] Audubon and Bachman, 1847: p. 235 (collared peccary); Leopold (1959), 1972: p. 496 (collared peccary); Alston, 1879–1882: p. 110 (white-lipped peccary).

[16] Bancroft, 1769: p. 125; Sánchez Labrador (ca. 1766), 1910: 1: p. 195; Wafer (1680–1688), 1934: p. 64.

[17] Fernández de Oviedo y Valdés (1526), 1950: p. 152, (1520–1555), 1959: 2: p. 45. Cieza de León ([1532–1550], 1853: p. 400, 1864: p. 174) and Hernández ([1571–1576], 1959: 1: pp. 311–312) knew that it was not a true navel ("pero no es un verdadero ombligo"), but the misunderstanding—or at least the description—persisted.

COYÁMETL O QUAUHCOYÁMETL

FIG. 1. (*Continued*)

powerful and unpleasant in the case of the white-lipped peccary. Again both peccaries, but notably the larger species, have bristle-like hairs on the neck and back ("mane"), whence, presumably, the description *puerco espín* or *puerco espino*,[18] also applied to the porcupine.[19] Sahagún compared the bristles of the peccary to awls,[20] Azara to the quills of the porcupine itself, *Coendou prehensilis*.[21]

---

[18] Montero de Miranda (1574), 1953: p. 348; Cobo (1653), 1956: 1: p. 358; Carrion (1581), 1965: p. 34; Jiménez de la Espada (ed.) (1594), 1965: 2 (2): p. 84. Cf. Hernández de Alba, 1948b: p. 337.

[19] Pennington (1979–1980: 2: p. 67) has "jabalí. puerco espín. *tasicori* [Névome or Pima Bajo]," apparently both peccary and porcupine (cf. ibid., 1: pp. 202, 211, 2: p. 98).

[20] Sahagún, 1963: p. 10; "cerdas ásperas" in Fernández de Oviedo y Valdés (1520–1555), 1959: 2: p. 45.

[21] Azara (1801), 1838: 1: p. 166 ("couiy," incorrectly *Hystrix prehensilis*); also R. F. Burton (ed.) in Staden, 1874: p. 160 ("porcupine quilled"). W. H. Smyth's translation (1857: p. 115) of Girolamo Benzoni's *La historia del mondo nuovo* has "bristles along the back." However the Italian (1572: p. 79) reads "il bellico [navel] sopra la schiena [spine]."

B: SCIENTIFIC NOMENCLATURE

Order, Artiodactyla (Owen, 1848).

The peccaries belong to the mammalian order Artiodactyla (previously a suborder of the Ungulata[1]), even-toed ungulates, to which important domestic animals (cattle, sheep, goats, camels, llamas and pigs) also belong.

Family, Dicotylidae.

Between the order Artiodactyla and the family Dicotylidae (or Tayassuidae) categories of suborder or superfamily (Suiformes,[2] Choeromorpha,[3] Suoidea[4]) are sometimes recognized. A minority of modern authorities have described the Dicotylidae—alternatively Dicotyles or Dicotylinae—as a subfamily under the Suidae.[5] The term Dicotylidae (from the Greek *dis*, "double" and *kotule*, "cavity"—"double navel") was introduced, as Dicotylina, by H. N. Turner in 1849,[6] the synonym Tayassuidae (from the genus *Tayassu*) by T. S. Palmer in 1897.[7] The former therefore takes precedence.

Genera, *Tayassu, Dicotyles, Catagonus.*

The relationships between the members of the Dicotylidae, and thus their respective biological status, have not been finally determined. The evidence presently available suggests the existence of three living genera. From the sixteenth century, several "kinds" (*castas, especies, generos*) of peccary have been reported. Linnaeus (1758) placed them all, together with the Old World pigs, in the genus *Sus*,[8] although his *Sus tajacu*

---

[1] Alston, 1879–1882: p. 106; Miller and Rehn, 1902: p. 12.

[2] Miller and Kellogg, 1955: p. 792; Soukup, 1961: p. 325.

[3] Cabrera, 1960: p. 315.

[4] Miller and Kellogg, 1955: p. 792; Sowls, 1974: p. 153.

[5] Alston, 1879–1882: p. 106; Trouessart, 1897–1899: 2: pp. 816–818, 1904: p. 658 (Tayassinae = Dicotylinae); Lydekker and Blaine, 1913–1916: 4: p. 374; Frenchkop, 1955: p. 526.

[6] H. N. Turner, 1849: p. 159.

[7] Palmer, 1897: p. 174.

[8] Linnaeus, 1758–1759: 1: p. 50, 1767–1770: 1: p. 103, [Gmelin], 1788–1793: 1: p. 219. Linnaeus (1758) also gives *Aper* ("wild boar") *Mexicanus Moschiferus*, following earlier authorities (Tyson, 1683: p. 359; Ray, 1693: p. 97; Klein, 1751: p. 25; Brisson, 1756: p. 111). Barrère (1741: pp. 160–162), writing of Cayenne, associated two peccaries (*tajacu* and *patyra*) with *Sus*.

referred specifically to the collared peccary (*Tayassu tajacu*).[9] *Sus* continued to be employed by some authors until about 1820.[10]

Most systematic accounts of the peccary place the species either (a) under one or other of two synonymous generic names, *Tayassu* and *Dicotyles*, or (b) under both where two genera are recognized. *Tayassu* was established by G. Fischer (1814),[11] *Dicotyles* by G. Cuvier (1817)[12]; each proposed a single genus to accommodate the two specific forms described from Paraguay by Félix de Azara (1801). Fischer later (1817) attempted to replace *Tayassu* with *Notophorus*.[13] The latter cognomen was adopted by J. E. Gray (1868) for the collared peccary (*N. torquatus*), *Dicotyles* being retained for the white-lipped peccary (*labiatus*).[14] Two other generic names have been proposed: *Adenonotus* (1828),[15] including both species, and *Pecari* (1835),[16] for the collared peccary (*torquatus*) alone. In 1901 C. H. Merriam introduced the subgenus *Olidosus* (under *Tayassu*) for the white-lipped peccary (*albirostris*).[17]

In the nineteenth century a single genus (generally *Dicotyles*) was preferred by the great majority of authors, and in the twentieth century either one (generally *Tayassu*) or two. M. O. Woodburne (1968), on the basis of a study of cranial myology and osteology, argued for two,[18] and this usage was accepted by A. M. Husson in his monumental work on the mammals of Surinam (1978). Husson further showed that *Tayassu* (Fischer, 1814) must be reserved for the genus containing the collared peccary, and *Dicotyles* (G. Cuvier, 1817) for that containing the white-lipped peccary.[19] However, the question of whether to recognize two genera is not finally settled. Recent work on a third peccary, assigned to

---

[9] Alston, 1879–1882: p. 107 n; Husson, 1978: p. 348.

[10] Erxleven, 1777: p. 185 (*Sus tajassu*); Zimmermann, 1777: p. 550 (*Sus tejacu*); Kerr [Gmelin/Linnaeus], 1792: pp. 352–353 (*Sus tajassu, S. tajassu minor, S. tajassu patira*); Link, 1794–1801: 2: pp. 104–105 (*Sus pecari, S. patira*); Humboldt (1811), 1966: 3: p. 51 (*Sus tajassu*); Illeger (1811), 1815: pp. 108, 115 (*Sus tajassu, S. albirostris*); Bowdich, 1821: p. 71 (*Sus pecaris, S. dicotylus*). Schomburgk (1837: p. 321) identified "the large peccary or Indian hog" of Guiana as *Sus cystiferous major*.

[11] Fischer, 1814: pp. 284–287 (*Tayassu pecari, T. patira*). Frisch's (1775: p. 3) *Tagassu*, followed by Elliot (1904: p. 61) and Goodwin (1946: p. 446), is non-Linnaean and unacceptable, as pointed out by Hershkovitz (1948: pp. 272, 274) and Woodburne (1968: p. 34 n.1). Trouessart (1904: p. 658) and Neumann (1967: p. 122) employ the latinized forms *Tayassus* and *Tagassus* respectively.

[12] G. L. C. F. D. Cuvier, 1817: 1: pp. 237–238. *Dicotyle* in Gervais and Ameghino, 1880: pp. 110–112.

[13] Fischer, 1817: p. 373.

[14] Gray, 1868: pp. 43–45.

[15] Brookes, 1828: p. 76.

[16] Reichenbach, 1835: part 6: p. 1. Followed by E. A. Goldman, 1920: p. 72; Gidley, 1920: pp. 655–656; Miller, 1930: p. 18; Murie, 1935: p. 28; Bertoni, 1939: p. 8; Cabrera and Yepes, 1940: p. 280; Hershkovitz, 1951: p. 566; Miller and Kellogg, 1955: p. 792.

[17] Merriam, 1901a: p. 120. Followed by Elliot, 1904: pp. 65–66; raised to generic status by Trouessart, 1904: p. 658.

[18] Woodburne, 1968: pp. 1–48.

[19] Husson, 1978: p. 348. Similarly identified in Bourlière, 1955b: p. 58.

the genus *Catagonus*, has suggested a "closer relationship" between the two well known species than was concluded by Woodburne.[20]

In the mid-1970s a living representative of the genus *Catagonus* (previously known only in fossil form, along with *Platygonus*, *Mylohyus*, and *Prosthenops*) was discovered in the Paraguayan Chaco.[21] It is now apparent that the territorial range of *C. wagneri* (with both Guaraní and Spanish names) extends to neighboring parts of Bolivia and Argentina.[22] According to R. M. Wetzel, "The three species of living peccaries meet in the Gran Chaco during what may be an interim period—the present— between a more arid cycle that favored thorn forest, steppe and *C. wagneri* and a moist cycle that will increasingly favor more mesic forests and *Tayassu*"[23] (the collared and white-lipped peccary).

Species (*pecari, tajacu*) and subspecies.

The earliest valid specific name for the white-lipped peccary is *pecari* (Link, 1795),[24] taking precedence over *albirostris* (Illiger, 1811) and *labiatus* (G. Cuvier, 1817). As to the collared peccary, *tajacu* (Linnaeus, 1758) is indisputably correct. The number of subspecies of *D. pecari* and of *T. tajacu* remain to be determined. C. A. Hill (1966) gives five and fourteen respectively,[25] probably conservative estimates. E. R. Hall and K. R. Kelson (1959) mapped the approximate distribution of ten subspecies of *T. tajacu* north of Panama.[26]

---

[20] Wetzel, 1977a: p. 7. C. Groves (1981: p. 2) and S. J. Olsen (1982: p. 9) place both species under *Tayassu*. Earlier, Villa (1948: p. 523) had argued that the difference between the two was less than generic. A. R. Wallace (1853: p. 450) and E. Liais (1872: pp. 402–403) suggested three "species," but all were placed within the genus *Dicotyles*.

[21] Wetzel, Dubos, Martin and Myers, 1975: pp. 379–381.

[22] Olreg, Ojeda and Barquez, 1976: pp. 53–56; Wetzel, 1977b: pp. 536–544; Mares, Ojeda and Kosco, 1981: p. 199.

[23] Wetzel, 1977a: p. 35.

[24] Osgood, 1921: p. 39; Hershkovitz, 1963: p. 86; Husson, 1978: p. 348.

[25] C. A. Hill, 1966: p. 6.

[26] Hall and Kelson, 1959: p. 996.

C: DISTRIBUTION

The Dicotylidae originated in North America,[1] but the living genera *Tayassu, Dicotyles* and *Catagonus* probably all evolved south of the Isthmus of Panama. Fossil *Tayassu* (collared peccary) and *Catagonus* have been reported from the Pliocene and Pleistocene of Argentina.[2] The earliest evidence of fossil (collared) peccary in Middle America belongs to the late Pleistocene of Guatemala.[3]

The distributions of the two most common species of living peccary (map 1) together cover the whole of Neotropica (*D. pecari* and *T. tajacu*) and adjoining parts of Nearctica (*T. tajacu*). Habitats range from semi-desert scrub to climax rain forest. Races of collared peccary are found from the southern United States (Arizona, New Mexico, Texas) to northern Argentina, excluding much of north-central Mexico, the Andean cordillera, and the Pacific littoral south of Ecuador. The white-lipped peccary belongs to the hot and humid tropics, from southern Vera Cruz (Mexico) to Paraguay. The most obvious question concerns their respective latitudinal limits.

Northern limits (map 2)

The collared peccary is now absent or very rare over a large part of the high plateaus of north-central Mexico.[4] From broad eastern and western salients, penetration of the central region is by deep canyons, as

---

[1] Burt, 1949: p. 216; Hershkovitz, 1972: p. 362. For *Platygonus* (Upper Pliocene and Pleistocene), see Leidy, 1853: p. 9; Trouessart, 1897–1898: 2: p. 816; Gidley, 1920: p. 651; Gazin, 1938: pp. 41–49; for *Mylohyus* (Pleistocene), see Gidley, 1920: p. 651; Lundelius, 1960: pp. 1–40; also, generally, Lavolat, 1955: pp. 547–549. Walker (1964: 2: p. 1366) and Cooke and Wilkinson (1978: pp. 476–477) give the geological range of the family (Tayassuidae or Dicotylidae) as: lower Oligocene to upper Miocene in Europe, lower Pliocene in Asia, Pleistocene to Recent in South America, and lower Oligocene to Recent in North America. Hendey (1976: pp. 787–789) reported the first fossil peccary (*Pecarichoerus africanus*) from the Pliocene of South Africa (Cape Province).

[2] Fossil Tayassuidae (Dicotylidae) were noted by Burmeister, 1876–1879: 3: p. 473, under *Dicotyles torquatus* (collared peccary), from a site within Buenos Aires. See also Gervais and Ameghino, 1880: pp. 110–113; Ameghino, 1889: 1: p. 574; Rusconi, 1931b: pp. 228–240, 1931a: pp. 121–227, especially pp. 196–197. According to Sowls (1969: p. 219), *Platygonus* (now extinct) and the collared peccary apparently lived together in parts of the range of the latter.

[3] Woodburne, 1969: pp. 121–125. Claims to the contrary notwithstanding (Leidy, 1853: p. 9; Trouessart, 1897–1898: 2: pp. 817–818), there is nothing from Mexico or the United States (Hershkovitz, 1972: p. 362).

[4] Maps 2 and 3 are based on site reports, regional references, and published sketch maps (Hall and Kelson, 1959: 2: pp. 996–997; Dalquest, 1949: p. 413; Leopold (1959), 1972: pp. 495, 499).

MAP 1. Approximate present distribution of peccaries.

in the Tepehuan country of Durango.[5] The "wild pig" is mentioned by Felipp Ségesser in his account of the Pimería (Tecoripa) in 1737.[6] It was important to the prehistoric Seri of Sonora.[7] The Huichol and Cora of Jalisco and Nayarit,[8] and the Cáhita (Mayo and Yaqui) of Sonora and Sinaloa continue to hunt the species.[9] The western distribution extends into central Arizona[10] and southwestern New Mexico.[11] To the east,

---

[5] Pennington, 1969: pp. 70, 128, 146. See also Anderson, 1972: pp. 389–390.

[6] Treutlein, 1945: p. 186.

[7] Bowen, 1976: p. 20.

[8] Grimes and Hinton, 1969: p. 797.

[9] Beals, 1943: p. 13, 1945a: pp. 10–13.

[10] McCullough, 1955: pp. 146–149; Neal, 1959: pp. 177–190; Eddy, 1961: pp. 248–257; Olsen, 1964: pp. 23–24, 157; Schweinsberg, 1971: pp. 455–460.

[11] Henderson and Harrington, 1914: p. 32; Bailey, 1931: pp. 10–11; W. J. Hamilton, 1939: pp. 153, 368.

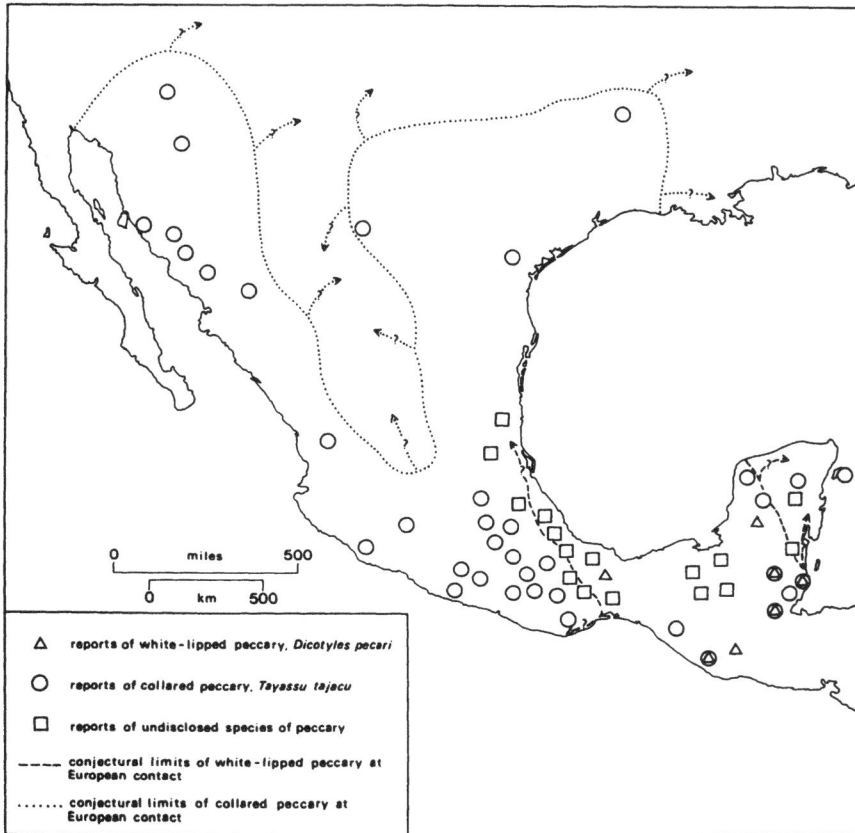

MAP 2. Northern limits of the peccary.

Tayassu tajacu is found in the scrublands of San Luis Potosí,[12] and of Coahuila and Tamaulipas, and thence into Texas[13] and extreme southeastern New Mexico.[14] Between a hundred and a hundred and fifty years ago it was known as far east as the Red River valley in Arkansas.[15] The destruction of natural habitat (pine-oak woodland as well as brushwood) and the introduction of firearms everywhere led to a contraction in the distribution of the peccary,[16] but within the United States the enforcement

---

[12] Dalquest, 1953: pp. 206–208; Laughlin, 1969: p. 300 (Haustec); Stresser-Pean, 1952–1953: p. 222 (Haustec).

[13] J. A. Allen, 1896: p. 54; W. J. Hamilton, 1939: pp. 153, 168; Dalquest, 1953: p. 207; Ellison and Harwell, 1969: p. 426; Anderson, 1972: pp. 389–390. Faunal evidence from Tamaulipas in MacNeish, 1958: p. 140 (Preclassic).

[14] Bailey, 1931: p. 10.

[15] Audubon and Bachman, 1847: p. 240; Baird, 1859: p. 627; Trouessart, 1904: p. 817 (ab Arkansas); Hamilton, 1939: p. 368; C. A. Hill, 1966: p. 67. The "western limit is not well ascertained, though it is said by some to occur in California" (Baird, 1859: p. 627, unconfirmed).

[16] According to D. D. Brand (1951: p. 162), the peccary became extinct in the *muncipio* of Quiroga on Lake Patzcuaro (Michoacán) within the last 50 to 100 years. For contraction in distribution elsewhere, see Hunn, 1977: p. 226 (Tenejapa, Chiapas, Mexico).

of game laws has been followed by some expansion in both range and numbers.[17]

The northern limit of the white-lipped peccary lies in southern Mexico, where the species is now rare or has locally disappeared in the course of the present century. There is only one known report from Campeche (1901).[18] Peccaries reported from northern and eastern Yucatán and northern Belize are either collared or of undisclosed species (probably collared).[19] A history of retreat southward and westward in the peninsula is indicated. However, the greater part of the Petén (Guatemala) still remains within the zone of *Dicotyles pecari*. In 1935 it was said to be the more common species around Uaxactun.[20] There are also reports from neighboring northern Vera Paz,[21] where Montero de Miranda (1574) noted the presence of two kinds of peccary,[22] and from northeastern Chiapas, the domain of the Chol-Lacandon.[23] Up to about a century ago, the species could still be found on the forested slopes of the Volcán de Atitlán.[24] Beyond Chiapas and, presumably, Tabasco, the distribution stretches to the Isthmus of Tehuantepec in southern Veracruz.[25] M. D. Coe and R. A. Diehl (1980) observed that the white-lipped peccary had been hunted to extinction in the vicinity of San Lorenzo Tenochtitlán.[26] There are several early notices of *puercos monteses* in the Vera Cruz region, as far north as Jalapa, and some may refer to *D. pecari*.[27] Probably

---

[17] Sowls, 1969: p. 223 (Arizona and Texas).

[18] Merriam, 1901b: pp. 120–121 (Apazote). Gaumer (1917: p. 67) lists *D. labiatus* (Cuvier), jabalí (*cehuikax*), among the mammals of Yucatán.

[19] The white-lipped peccary ("warree") was first reported from Belize (British Honduras, no precise location) by R. Temple, 1860: pp. 206–207. Leopold ([1959], 1972: pp. 498, 499) includes the whole of Yucatán within the range of the white-lipped peccary. Tusks of both species have been found at Dzibilchaltun and Mayapan, northern Yucatán (Pollock and Ray, 1957: p. 639; Pollock, Roys, and Proskouriakov, 1962: p. 377; Wing and Steadman, 1980: pp. 326–327), but the animals may not have been captured or killed locally. Not reported archaeologically from Cozumel (Hamblin, 1980: p. 223; 1984: pp 123, 126).

[20] Murie, 1935: p. 28. Faunal remains of white-lipped peccary were reported by Ricketson and Ricketson, 1937: p. 204, and Kidder, 1947: p. 60; similarly from Altar de Sacrificios by Olsen, 1972: p. 244 (Classic and early Postclassic) and from Altar, Seibal, Macanche and Tikal by Pohl, 1976: pp. 97 ff. Both species are "now abundant when ramon nuts fruiting" (Pohl, 1976: p. 59).

[21] Alston, 1879–1882: p. 109.

[22] Montero de Miranda (1574), 1953: p. 348. Cf. Ximénez (1722), 1967: pp. 57–58. Stoll (1889: p. 25), Brigham ([1887], 1965: p. 370) and Ibarra (1959: p. 171) refer to the white-lipped peccary in Guatemala.

[23] Alvarez del Toro, 1952: p. 192 (*T. pecari*, "senso"); Anon., 1976: p. 29 (*T. pecari*, "tamborcillo," "senso"). Cf. Palacios, 1928: p. 150; Duby and Blom, 1969: p. 281; Hellmuth, 1977: p. 430 [1695].

[24] Alston, 1879–1882: p. 110. The site of Kaminaljuyú has yielded faunal remains (Kidder, Jennings and Shock, 1946: p. 157).

[25] Dalquest, 1949: pp. 411–413 (17. 30 N., reports to 18. 30 N.), including distribution map; Hall and Dalquest, 1963: pp. 352–353 (as far north as the Tuxtla mountains). Cf. Garay, 1846: p. 67; J. J. Williams, 1852: pp. 205–207; Gadow (1902–1904), 1908: p. 374 (*Dicotyles labiatus*, "jabali de manado," "moro" ["moreno"]); Foster, 1942: p. 26 (Popoluca territory).

[26] Coe and Diehl, 1980: 2: p. 102.

[27] Paso y Troncoso (ed.), 1905b: pp. 104 [1580], 107 [1580], 111 [1580]; ibid: p. 199, and Latorre (ed.), 1920: p. 50 [1571]. See also Ajofrín (ca. 1765), 1958–1959: 2: p. 29 (Cotaxtla).

the range also included the neighboring forested foothills of northern Oaxaca.[28] To the south, the species apparently survives along the Pacific coast of Guatemala[29] and probably at one time extended into Soconusco (southern Chiapas).[30]

Southern limits (map 3)

The distribution of peccaries stretches southward from the Guiana-Brazilia heartland, between the eastern flanks of the Andes[31] and the Atlantic seaboard. They were formerly present throughout the Gran Chaco and thence along the eastern piedmont and down the valleys draining to La Plata. The distribution has probably steadily contracted since the beginning of European settlement, particularly in the vicinity of major centers of population.[32] In the territory of the Tsátchela (Colorado) Indians of western Ecuador the peccary is said to have "gradually disappeared, due to a sort of murrain, introduced with horses by the white settler."[33]

Florentino Ameghino (1889) maintained that, in pre-Columbian times, the collared peccary ranged as far as the province of Río Negro ("north of Chubut," that is about latitude 42 degrees).[34] Other authorities put the southern limit (historically) between the Río de la Plata and the Río Negro.[35] Sixteenth-century *relaciones* of Asunción and Tucumán mention "puercos del monte"[36] and "puercos jabalies."[37] The collared peccary is

---

[28] Paso y Troncoso (ed.), 1905a: p. 66 (Chinantla); Fuente, 1947: pp. 167, 199 (Choapan, Comaltepec); Ajofrín (ca. 1765), 1958–1959: 2: pp. 65 (near Teutila/Jalapa de Díaz), 70 (Zoyaltepec).

[29] Alston, 1879–1882: pp. 109–110 (quoting Godman and Salvin); Coe, 1961: p. 12; Coe and Flannery, 1967: p. 117.

[30] Pineda (1845: p. 24) mentions "javali" (*censo* or *senso*, usually the white-lipped peccary) in Chiapas and Soconusco. Villa (1948: p. 423), in his study of the mammals of Soconusco, gives only *Tayassu angulatus* (collared peccary).

[31] No evidence of peccaries in lands to the south of Ecuador and west of the Andes has been found. J. I. Molina ([1782] 1809: 1: p. 223) made the obscure comment: "I do not consider [the dog and the hog in Chile] as proceeding from a European stock, as the proper names which they both have in the Chilean language distinguished them from foreign animals." The Mapuche (Araucanian) words for "jabalí," *cùchrecùchre* and *sañhue* (Erize, 1960: p. 498), presumably refer to feral pigs. See also Plagemann, 1888: p. 315 ("eine Art jagdbarer Wildschweine").

[32] Belaieff (1946: pp. 373–374) remarked on the decline of the Chaco peccary, and added, "and no kind of valuable game lives within thirty leagues of the confluence of the Paraguay and Pilcomayo rivers" (that is, around Asunción).

[33] Von Hagen, 1939: p. 5 (no authority given, and no similar report from elsewhere has come to light).

[34] Ameghino, 1889: 1: p. 574. Audubon and Bachman (1847: p. 240, on the authority of Azara) give 37 degrees south. E. R. Alston (1879–1882: p. 107), L. S. Crandall (1964: p. 528) and E. P. Walker (1964: p. 1365) refer to peccaries in "Patagonia" (at least as far as the Río Negro).

[35] Darlington, 1957: p. 403; Gilmore, 1963: p. 382. Cf. Lozano (ca. 1730–1745), 1873–1874: 1: p. 286.

[36] Jiménez de la Espada (ed.), 1965: 2 (2): pp. 80 (1594), 84.

[37] Latorre (ed.), 1919: p. 144.

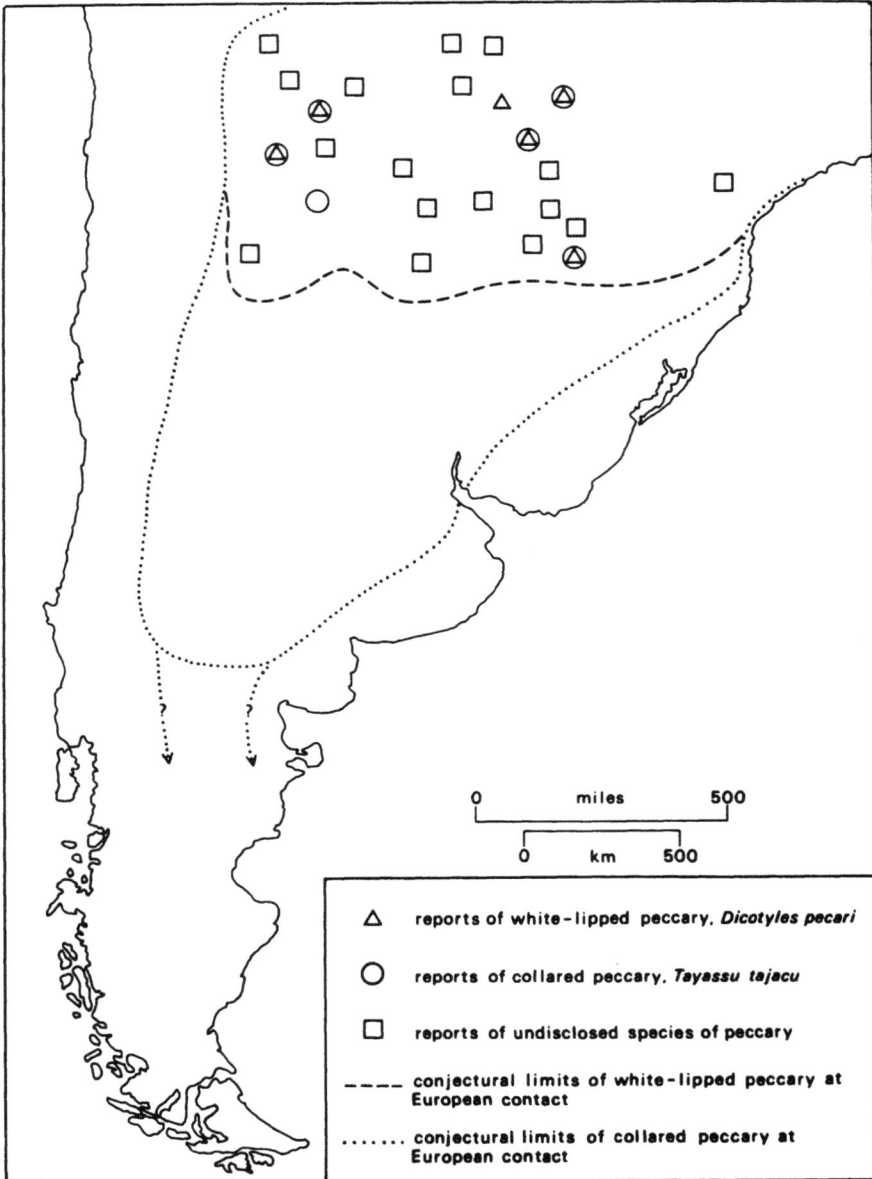

MAP 3. Southern limits of the peccary.

still common in the flatlands and hill country of the eastern part of the
province of Salta "in sympatry with *Catagonus wagneri*," the recently
discovered third living species of peccary.[38] Pedro Lozano (ca. 1730–

---

[38] Mares, Ojeda and Kosco, 1981: p. 199.

1745) appears to refer to both of the principal species in the Chaco,[39] and to one or other (probably *T. tajacu*) in writing of the Guayakí of southern Paraguay.[40] Peccary (*T. tajacu*, if not also *D. pecari*) have been reported from the Chaco in territories occupied by the Chamacoco,[41] the Guaná or Chaná,[42] the Terena,[43] the Chorotí,[44] the Mataco,[45] the Pilagá,[46] and the western Toba,[47] and in southern and eastern Paraguay, and neighboring Misiones (Argentina), by the Mbyá, the Chiripá[48] and the Cainguá.[49] The most southerly known report from Brazil concerns the Caingang.[50]

The white-lipped peccary formerly ranged as far south as northern Argentina (El Chaco to Misiones[51]), probably just beyond the middle Paraná. E. Boman (1908) implies the presence of both species in the Sierra Santa Barbara, Jujuy, Argentina.[52] The western Toba hunted the white-lipped peccary along the forested margins of the Bolivian and Argentinian Chaco.[53] The missionary Martin Dobrizhoffer (ca. 1780) reported "four different species of wild boar," including perhaps the white-lipped peccary, in his account of the Abipón in the southern Chaco, west of the lower Paraná.[54] Probably the whole of Paraguay and neighboring parts of Argentina lay within the range of the larger species at the time of first European contact.[55]

Off-shore islands

Peccaries have been recorded for a number of off-shore islands. When and in what circumstances they were introduced are unknown. Grijalva

---

[39] Lozano, 1733: pp. 37, 40 (Gran Chaco Gualamba).

[40] Lozano (ca. 1730–1745), 1873–1874: 1: p. 415 (Guachaguís). For the peccary among the Guayakí, see also Vellard, 1939: p. 88; Métraux and Baldus, 1947: p. 438; Cadogan and Colleville, 1963: p. 442; Cadogan, 1973: p. 98.

[41] Métraux, 1946a: p. 302.

[42] Sánchez Labrador (ca. 1766), 1910: 1: p. 258; and cf. ibid: pp. 194–196.

[43] Oberg, 1949: p. 10; and, to the west, the Caduveo (ibid: p. 59).

[44] Nordenskiöld, 1912: p. 50; Karsten, 1932: pp. 38–39. See also Nordenskiöld, 1920: p. 31 (Chané and Chiriguano).

[45] Nino, 1913: p. 97 (Guisnay = Mataco); Métraux, 1946a: pp. 261, 264.

[46] Métraux, 1946a: p. 264.

[47] Métraux, 1946a: p. 261.

[48] Cadogan, 1973: p. 98.

[49] Ambrosetti, 1894a: pp. 679, 702, 726, 729.

[50] Métraux, 1946b: p. 451; Henry (1941), 1964: pp. 100, 157.

[51] Ameghino, 1889: 1: p. 574; Ambrosetti, 1894b: p. 68.

[52] Boman, 1908: 1: pp. 90–91.

[53] Karsten, 1923: p. 116, 1932: pp. 4, 38–39.

[54] Dobrizhoffer (first published, in Latin, Vienna, 1784), 1822: 1: pp. 89, 270.

[55] Félix de Azara (ca. 1800), who first distinguished between the species, has little to say about their respective distributions, except that both occupied territory to the north of the Río de la Plata (1809: 1: p. 249; repeated by Beaumont, 1828: p. 32). See also Rengger (1818–1826), 1830: p. 322, 1835: pp. 207–208 (Paraguay); Métraux, 1946a: pp. 257, 336 (the Mbayá, Paraguay-Brazil); Regehr, 1979: p. 28 (Paraguayan Chaco). Other southerly reports of the white-lipped peccary in O. Thomas, 1903b: p. 242; Miller, 1930: p. 18 (southern Mato Grosso).

and Cortés (1518–1519) found "puercos monteses" on the Isla de Cozumel,[56] eighteen kilometers from the eastern coast of Yucatán. The first modern report was by C. H. Merriam (1901), who claimed the specimen as a new species, *Tayassu nanus,* only two-thirds the size of the common collared peccary (possibly the result of long isolation).[57] This is supported by recent archaeological work.[58]

There is prehistoric evidence of *Tayassu* sp. in Trinidad,[59] as well as modern scientific references.[60] Thomas Jeffreys (1762) found "plenty of wild hogs,"[61] but he may have been referring to feral pigs. In Tobago, some forty kilometers northeast of Trinidad, there was, according to César Rochefort (1666), "une sorte de sangliers, que quelques Indiens nomment *javaris* et les autres *pequires.*"[62] About 150 kilometers to the northwest, on the island of Carriacou, a clay head "resembling . . . the peccary" has been found.[63] Bryan Edwards (1793) maintained that the peccary was "anciently" present in the Windward Islands (thus "proving" that they had been peopled from the south), but had subsequently been exterminated. The specimens that he had actually seen had been "carried thither from the continent as objects of curiosity."[64] Some may have been taken as far afield as the Leewards.[65]

J. E. S. Linné (1929) suggested that peccaries "were kept as domestic animals on the Pearl Islands [thirty kilometers off the Pacific coast of Panama], at all events during the later [prehistoric] period. . . ."[66]

---

[56] López de Gómara (1551–1552), 1954: 2: 28. *Puercos monteses* usually refers to the peccary. Fernández de Oviedo y Valdés ([1526], 1950: p. 151) also applied the description to feral European pigs on the West Indian islands. However, Díaz del Castillo's ([ca. 1568], 1955: 1: p. 62 [1518]) "muchos puercos de la tierra [Cozumel], que tienen sobre el espinazo el ombligo" clearly refers to the peccary.

[57] Merriam, 1901a: p. 102. On the peccaries of Cozumel, see also Gaumer, 1917: p. 65; Hershkovitz, 1951: p. 567; Vos, Manville and Gelder, 1956: pp. 174, 176; Jones and Lawlor, 1965: p. 417.

[58] Hamblin, 1980: p. 234.

[59] Wing, 1977: p. 58. Trinidad is about sixteen kilometers from the shores of Venezuela.

[60] Dorst, 1967: p. 39 (*D. tajacu*); Méndez, 1970: p. 243 (*T. tajacu*). Kerr ([Gmelin/Linnaeus], 1792: p. 352) observed that "[*Sus tajassu*] inhabits the warmest parts of America and . . . some of the West India islands."

[61] Jeffreys, 1762: p. 3 (cf., ibid.: p. 17, "wild boars called *sajones*" around Cathagena [Cartagena], probably peccaries).

[62] Rochefort, 1666: p. 31. Also described in Rochefort and Poincy, 1658: p. 122, 1666: p. 70 (other islands too?).

[63] Fewkes, 1922: p. 121.

[64] Edwards, 1793: 1: pp. 87, 89. The Windwards extend from Dominica in the north to Grenada in the south and include Carriacou. D. Taylor (1938: p. 149) observed that "the Caribs [of Dominica] . . . rely upon their dogs to find and kill agouti and [wild] pig." However the latter were probably feral *Sus scrofa.*

[65] Roulin, 1835: p. 329 ("[J.-B. du Tertre] connaissait fort bien [les pécaris] qui, de son temps [ca. 1650], étaient quelquefois apportés de la côte de Cumana à Saint-Christophe par les barques venant de l'île de Tabago"). Martius (1867: 2: p. 318) gives *zaino scuna* as the Taino (Española) name for the peccary. For *zaino* (Spanish) see pp. 50–51 infra.

[66] Linné, 1929: pp. 130–131.

However, the species has never been described (archeologically or otherwise) from these islands or from the neighboring island of Coiba.[67]

Collared peccaries are said to have been introduced to Saint Domingue (Haiti) and the Ile de la Gonave from Cartagena some time before 1787 by the then governor of the islands, La Luzerne.[68] The idea of naturalizing the species, and thereby increasing the food supply, appears to have been thwarted by the Negro rebellion.

---

[67] See Bangs, 1901: pp. 631–644; O. Thomas, 1902: pp. 135–137, 1903a: pp. 34–42; Thayer and Bangs, 1905: pp. 135–160; Kellogg, 1946: pp. 1–4.

[68] M. L. E. Moreau-Saint-Méry (trans.) in Azara, 1801: 1: p. 42; Audubon and Bachman, 1847: p. 239; Armas, 1888: p. 170.

D: HABITAT AND DIET

Habitat

The peccaries, like the wild pigs of the Old World, belong to a variety of woody environments and subsist on the products of the forest and of woodland and scrub formations. Of the two most common New World species, the collared peccary has much the greater latitudinal and environmental range, extending from the hot and humid tropics to lands that are seasonally colder and drier, and less heavily vegetated, than those occupied by the white-lipped peccary. The greater environmental tolerance of *Tayassu tajacu* is apparently matched by a significantly larger number of subspecies or varieties.

There is a considerable area of overlap in the general distribution of the two species (map 1), which, according to P. Hershkovitz, "occupy similar if not practically identical niches in most of the Brazilian Subregion [of Neotropica] where they are sympatric."[1] Nevertheless there are differences in preferred local habitat even within the humid tropics, and the two species do not consort together.[2] There are no known reports of natural hybrids, although interbreeding has occurred in conditions of captivity.[3]

White-lipped peccary: The white-lipped peccary is the less closely observed, as well as the more conservative, species, and "further information regarding its actual habitat is desirable."[4] Moving in large, compact groups, it prefers climax rain forest, with a dense canopy and comparatively little undergrowth.[5] In mythology and folklore, the species and its environment are often closely associated.

The largest numbers (and larger herds) probably inhabit undisturbed sections of riverine and marshy lowland, with much smaller numbers in interfluvial tracts (*tierra firme*) and cooler montane forest. In some cases the latter may be relic or fugitive populations, following human exploitation of the valley lands. There are reports of white-lipped peccaries at

---

[1] Hershkovitz, 1972: p. 363.
[2] Remarked by Azara (1801), 1838: p. 118.
[3] Gray, 1864: p. 43 (London Zoological Garden); Alston, 1879–1882: p. 107; Flower and Lydekker, 1891: p. 290; Zuckerman, 1953: p. 906; Crandall, 1964: p. 529 (this casts doubt on the argument in favor of differences of a generic order between the two peccaries). R. K. Enders (1935: p. 478: Panama Canal Zone) stated that when white-lipped peccaries move into an area, other animals (presumably including the collared peccary) move out.
[4] Husson, 1978: p. 249.
[5] Enders, 1935: pp. 477, 479–480; Leopold (1959), 1972: p. 498.

ca. 5,000 feet in Venezuela[6] and Panama[7] and on the upper slopes of the Volcán de Atitlán in Guatemala.[8] They apparently range (or ranged) up to 6,000 feet or more on the eastern slopes of the Andes, "climatically very different from . . . eastern Brazil and the Amazon valley, but the [white-lipped peccary's] general characters are not peculiar."[9] In the usually short dry season, "they concentrate along rivers and forest creeks and not seldom can they be seen swimming across rivers."[10] Probably the white-lipped peccary rarely moves far from running water or shallow pools, where it is sometimes found wallowing.[11] Both species appear to make use of a variety of microenvironments to combat extremes of temperature.[12]

Collared peccary: Collared peccaries prefer dense undergrowth of secondary or remnant forest, paludal thickets, and tracts of scrub savanna (in the Chaco and elsewhere). Moving in single file, they can force a way through apparently impenetrable vegetation and sometimes form regular tracks. Man-made trails are generally avoided. When flatlands are periodically flooded, herds congregate on temporary islands of higher ground.[13] On the other hand, collared peccaries have been observed at up to 8,000 feet in the *monte alto* of Central and lower Middle America.[14] In higher latitudes, they inhabit pine and scrub-oak woodland[15] and, along the northern extremities of their range, xerophytic brushwood (chaparral, cactus, mesquite, and acacia), notably along ravines and over rocky slopes, and avoiding the dry and open flatlands.[16] In Sonora and Chihuahua peccaries are found up to ca. 7,500 feet,[17] in central and southeastern Arizona to ca. 6,500 feet.[18] In these peripheral areas the availability of water, ground cover and food supplies, and low nocturnal

---

[6] Röhl, 1959: p. 141.

[7] Anthony, 1916: pp. 364–365; E. A. Goldman, 1920: p. 75.

[8] Alston, 1879–1882: p. 110 (also higher woodland in Costa Rica).

[9] Osgood, 1914a: p. 151. See also Soukup, 1961: pp. 325–326 (*montaña* of the province of Sandia, Peru, at ca. 3800 feet).

[10] Husson, 1978: p. 352 (Surinam, quoting Geijskes, 1954). In the Vaupés region (Colombia), white-lipped peccaries occupy higher ground during the wet season, feeding on umari fruits, and later move towards the swamps (Silverman-Cope, 1973: p. 69). Husson also refers (on the authority of Penard and Penard, 1905) to dry-season migration towards the coast. R. H. Schomburgk ([1840–1844], 1923: 2: p. 130) observed that the white-lipped peccary swims "awkwardly," P. Hershkovitz (1972: p. 363) that both species are "excellent swimmers and cross rivers routinely." Husson (1978: p. 355) had no evidence of swimming by collared peccaries.

[11] Schomburgk, 1837: p. 321; Smith, 1976: p. 456 (mud hollows).

[12] Enders, 1935: p. 472; Sowls, 1969: p. 222; Perry, 1970: p. 39.

[13] Wagley and Galvão, 1949: p. 57; Wagley, 1977: p. 62.

[14] Frantzius, 1869: p. 296 (Costa Rica); Alston, 1879–1882: p. 108 (Guatemala); Goodwin, 1946: p. 447 (Costa Rica); Kelly and Palerm, 1952: p. 74 (Vera Cruz, Mexico).

[15] Pennington, 1969: p. 128 (Chihuahua).

[16] Bailey, 1931: pp. 10–11 (northern Mexico, Texas); Dalquest, 1953: pp. 207–208 (San Luis Potosí); West, 1964: p. 369 (western Sonora, north-western Sinaloa).

[17] W. J. Hamilton, 1939: p. 368.

[18] Neal, 1959: p. 177.

temperatures limit the distribution; and, from time to time, winter cold and shortage of food take their toll of numbers.[19] L. K. Sowls found "an apparent direct relationship" between the amount of rainfall (and thus the quantity and quality of their food supply) and the percentage of young in samples of animals killed by hunters.[20]

Diet

Both peccaries are omnivorous, although predominantly vegetarian. Francisco Hernández (1571–1576) referred to roots (*raíces*), acorns "y otros frutos del monte," as well as to grubs and worms (*gusanos, lombrices*) "y otros animales semejantes que se crían en sitios húmedos, lacustres y pantanosos."[21] Such animals, found on or just below the surface of the ground, include many species of insects, toads and reptiles, and snakes. Again, as opportunity arises, peccaries will take the eggs of birds, turtles and alligators, fish trapped in pools, and even some carrion.[22]

Peccaries are notorious crop-robbers, operating particularly at night. Raids on maize *milpas* (toppling the stalks to consume the cobs[23]) have been reported from the time of Oviedo (1526).[24] Using their prominent tusks, they grub for sweet potatoes and manioc roots, both the sweet and bitter varieties.[25] Among the Quechua-speaking Canelo of eastern Ecuador, the peccary is known as *lumu cuchi*, the "manioc pig."[26] A taste for sugar cane and bananas has also been remarked.[27] Raids on subsistence crops may locally be tolerated (or even encouraged) as a means of concentrating supplies of animal protein, the "garden hunting" reported from Panama.[28] Similarly the Maracá (Colombia-Venezuela) most fre-

---

[19] Leopold (1959), 1972: p. 496; Perry, 1970: p. 39.

[20] Sowls, 1966: p. 170.

[21] Hernández (1571–1576), 1959: 1: p. 311. According to Bernardino de Sahagún ([ca. 1570], 1963: p. 10), "its food is acorns, American cherries, maize, roots, fruit, just like what a pig eats. Hence they call the peccary a pig."

[22] Audubon and Bachman, 1847: p. 237; Sowls, 1969: p. 220; Leopold (1959), 1972: p. 496; Silverman-Cope, 1973: pp. 276–277. Husson (1978: p. 352, quoting Penard and Penard on the white-lipped peccary in Surinam) also mentioned fish.

[23] Pennington, 1969: p. 70 (country of the Tepehua, Durango, Mexico).

[24] Fernández de Oviedo y Valdés, 1950: p. 94.

[25] Eckart, 1785: p. 512; Sánchez Labrador (ca. 1766), 1910: 2: p. 258 ("las raíces de batatas y mandioca," territory of the Chaná or Guaná, northern Paraguay); Chapman, 1929: p. 74 (Barro Colorado island, Panama Canal Zone); Le Roy Gordon, 1957: p. 72 (Sinú region, Colombia); Moser and Taylor, 1963: p. 444 (Río Piraparaná, extreme southwest Colombia, territory of the Tukano); Villas Boas, 1974: p. 265 (upper Rio Xingu, Brazil).

[26] Cabrera y Latorre and Yepes, 1940: p. 280; Orr and Wrisley, 1965: p. 20; Whitten, 1976: p. 42.

[27] Azara (1801), 1835: p. 119; Chapman, 1938a: p. 39.

[28] Linares, 1976a: pp. 331–349, especially pp. 338–339, 345 (archaeological and contemporary evidence from the Aguacate peninsula). Earlier, Sauer (1966: p. 244) had suggested that "game [deer and peccaries] was perhaps most numerous near the settlements [of the Cueva], by the attraction of growing crops and uncultivated fields."

quently hunt peccaries from stands (sikarto) erected in cultivated fields and along paths, with maize, plantains, and bananas placed so as to attract the animals.[29]

It is accepted that peccaries subsist chiefly on wild plants (fruit, nuts and seeds, rhizomes and bulbs, mushrooms and other fungi, and fresh greens), many of which, in the humid tropics, are eaten by both species. Perhaps most important are the fruit and nuts of species of palm,[30] such as *Astrocaryum vulgare* (awarra), *Jessenia polycarpa* (milpeso),[31] *Attalea regia* (maripa),[32] *Euterpe oleracea* (açaí), *Iriartea* sp. (paxiúba),[33] *Scheelea zonensis*,[34] and *S. liebmannii* (coyal, coyal real).[35] The fruits of cacao (*Theobroma cacao*),[36] zapote (several species), umari (*Poraqueiba sericea*),[37] and membrillo (*Gustavia superba*)[38] are also readily devoured. The white-lipped peccary is the better equipped to crack hard nuts.[39] Some occasionally pass through the alimentary system in unbroken form, thus serving to disseminate particular tree species.[40]

Seasonal changes of diet and of range according to the availability of resources are probably most important towards the drier margins of the distribution of (collared) peccaries.[41] This is also, however, a feature of the humid tropics. R. K. Enders, writing of Barro Colorado Island (Panama Canal Zone), has described shifts in reliance upon *almendro* nuts (? *Geoffraea superba*), followed by "[wild] figs of many species," and then palm nuts.[42] Exudate gums and resins are sometimes eaten,[43] and the flow of gum may be stimulated by the gnawing of bark by the peccaries. Species of grass are perhaps chiefly "famine foods" and are generally more typical of impoverished (periodically arid) environments.[44]

The resources of the drier and/or temperate lands of the collared peccary include cactus stems (also a source of water) and fruit, notably the prickly pear (*Opuntia* spp.), acorns and pine nuts, the berries of juniper and manzanita (*Arctostaphylos glauca, A. tomentosa*), wild pota-

---

[29] Ruddle, 1970: p. 41. According to Nietschmann (1973: p. 167 n.), at Tasbapauni (Miskito coast, Nicaragua) "collared peccaries are frequently killed in the plantations as they root up the plants, but they are usually left or dragged off into the bush" (the meat is rejected as food).

[30] Cuervo (ed.), 1891–1894: 2: pp. 317, 319 (provincia de Zitará y curso del Río Atrato, 1761–1789).

[31] West, 1957: p. 41 (Pacific lowlands of Colombia).

[32] Husson, 1978: p. 352 (Surinam).

[33] Smith, 1976: p. 456 (Brazil).

[34] Enders, 1935: pp. 471–472 (Panama).

[35] Coe and Diehl, 1980: 2: p. 102 (southern Vera Cruz, Mexico).

[36] Schomburgk, 1836: p. 269 (Essequibo region, Guyana).

[37] Silverman-Cope (1973: p. 69) gives *Geoffroya spinoza*, Usher (1974) *Geoffroea superba*.

[38] Enders, 1935: p. 472 (Panama).

[39] Kiltie, 1979: p. ii.

[40] Chapman, 1931: p. 349.

[41] Eddy, 1961: pp. 248, 251.

[42] Enders, 1935: pp. 471–472.

[43] Enders, 1935: p. 472.

[44] Leopold (1959), 1972: p. 496; Enders, 1935: pp. 472, 478.

toes,[45] and mesquite (*Prosopis juliflora, P. pubescens*) and acacia beans.[46] The collared peccary is apparently (and probably necessarily) the more efficient exploiter of bulbs and rhizomes.[47] It can locate such foodstuffs at a depth of five to eight centimeters[48] by using the sense of smell.[49] The size and range of herds are related to the territorial and seasonal distribution of food supplies, in particular the association between the characteristically large herds of white-lipped peccary and the presence of "clumped resources."[50] The larger herds of both species are to be found in ecologically richer environments. Peccaries are often said to feed "erratically," but regular feeding grounds (*comederos*) have been described from areas normally undisturbed by man.[51] They may also frequent salt licks, around mud holes and along streams.[52]

---

[45] W. J. Hamilton, 1939: p. 153.

[46] T. A. Eddy (1961: pp. 248–257), in one of the few studies of the foods and feeding habits of the (collared) peccary, lists 40 species (providing fruit, berries, acorns, beans, foliage, tubers and nuts) from three study areas in southern Arizona.

[47] Kiltie, 1979: p. ii.

[48] Walker, 1975: p. 1366.

[49] Eddy, 1961: p. 255.

[50] Kiltie, 1979: pp. ii, 112, 144.

[51] Allen and Barbour, 1923: p. 261.

[52] Santiago Bertoni, 1973: p. 66; Vickers, 1976: p. 102.

E: BIOLOGY AND BEHAVIOR

The collared peccary has been intensively studied along the northern (arid) margins of its distribution, particularly in Arizona. There are also some detailed observations of both species from Barro Colorado island in the Isthmus of Panama. The white-lipped peccary is generally less well known, and the extent and significance of any regional variations in the behavior of the more numerous varieties of collared peccary are at present difficult to establish.

Reproduction and mortality. Peccaries breed from the age of twelve to twenty months. They do not form long-term pair bonds. Litters are small by comparison with domestic pigs, usually two over a period of twelve months.[1] Peccaries are not, therefore, "very prolific" as maintained by some eighteenth-century authors.[2] The gestation period of the white-lipped peccary (158 days[3]) is generally longer than that of the collared peccary (142 to 145 days or a wider range[4]). Births may occur at any time, although there is some evidence of a peak in summer (wet season) in areas with a sharply contrasting climatic regime.[5] The young of the collared peccary move around after a few hours, are weaned after six to eight weeks, and remain with the mother for about a year.[6] Captive specimens of the same species have been known to live for twenty or more years, the white-lipped peccary for up to thirteen years.[7] In the natural state, peccaries form resilient populations, with relatively low mortality rates. Consequently they may "multiply rather quickly, even though females produce only two young per year."[8] Apart from man, their principal enemies are the jaguar, the puma,[9] and less commonly the boa (*Eunectes murinus*).[10]

---

[1] Rarely twice in the course of a year—Roots, 1966: p. 198 (white-lipped peccary); Lewis, 1970: p. 46 (collared peccary); Leopold (1959), 1972: p. 498.

[2] Buffon (ca. 1770), 1884: 9: p. 234; Goldsmith, 1774: 3: p. 188; Bewick, 1790: p. 135; Stedman, 1796: 1: pp. 355–356; Bolingbroke, 1807: p. 227.

[3] Roots, 1966: p. 198.

[4] Sowls, 1961: p. 426 (142–148), 1966: p. 161 (approximately 145), 1974: p. 158 (142–145); Walker, 1975: 2: p. 1366 (142–148). A. S. Leopold ([1959], 1972: p. 496) refers to records of 96, 112 and 116 days.

[5] Sowls, 1966: p. 158, 1974: pp. 144, 157; Leopold (1959), 1972: p. 496.

[6] Leopold (1959), 1972: p. 496; Sowls, 1974: p. 162. Cf. Walker, 1975: 2: p. 1366 (two to three months).

[7] Crandall, 1964: pp. 529–530; Walker, 1975: 2: p. 1366.

[8] Husson, 1978: p. 352 (quoting Geijskes, 1954, white-lipped peccary).

[9] Schomburgk (1840–1844), 1923: 2: p. 130; Ross in Lizot, 1979: p. 154 n. 2; Kiltie, 1980: p. 542.

[10] Kappler, 1887: p. 81. Nelson (1916: p. 448) adds the bobcat (*Lynx rufus*) and the coyote (*Canis latrans mearnsi*).

FIG. 2. A collared peccary, standing on a snake. Thomas Bewick, 1790: p. 134.

Size of herds. Both peccaries are gregarious. They form herds throughout their lifetime, with little or no tendency to separate into temporary, seasonal groups according to age or sex.[11] The size of the herd is probably broadly related to the density of animals within a particular region. Over time, it is of course also affected by the ratio of births to deaths and by the migration of single animals or pairs, or more rarely small groups, between one herd and another (of the same species).

Actual numbers vary considerably, but herds of white-lipped peccary, the *jabalí de manada,* are almost always the larger, normally much larger. Fifty is unremarkable, and two hundred or even more have been reported.[12] Collared peccaries, on the other hand, usually form groups of ten to twenty, rarely as many as fifty.[13] Isolated pairs and even solitary individuals (old males) have occasionally been observed.[14]

Territory. Peccaries may be found moving about at any hour of the day or night. Probably both species are predominantly diurnal; in hot regions

---

[11] Sowls, 1974: pp. 144, 147 (collared peccary). E. R. Alston (1879–1882: p. 110), quotes a suggestion of F. D. Godman and O. Salvin [pers. comm.] to the effect that "female [white-lipped peccaries] with their young broods probably [keep] apart from the herd until the latter are of a sufficient size to shift for themselves," but this suggestion has not been confirmed.

[12] Barrère, 1741: pp. 160–162 (up to 1000); Stedman, 1796: 1: pp. 355–356 (up to 300, Surinam); Anon. (1607), 1908: p. 151 (40 to 300, Panama); Latorre (ed.; 1585), 1919: p. 86 ("duzientos y mas"); Wafer (1680–1688), 1934: p. 102 (200 to 300, Darién); León Pinelo (1650), 1943: 2: p. 51 ("mas de trescientos"); Husson, 1978: p. 352 (200 to 300 "in virgin forests [of Surinam]", quoting Penard and Penard, 1905). R. Perry (1970: p. 45, quoting Up de Graff on the Río Napo) refers to "a herd, estimated to be 2000 strong, swimming across a river."

[13] Hall and Kelson, 1959: p. 955; Anderson and Jones, 1967: p. 389. Cf. Perry, 1970: p. 39 (400 to 500, near the Rio Fresco, Mato Grosso: ? possibly misidentified).

[14] Enders, 1935: p. 469; Leopold (1959), 1972: p. 494.

they feed chiefly in the early morning and late afternoon or early evening. From what little evidence there is, it appears that the white-lipped peccary shows the greater tendency to nocturnal activity.[15]

There is little information on the "territories" occupied by herds of white-lipped peccary. Presumably the size of such areas is directly related to the herd's substantial food requirements and, in turn, to the resources of particular regions. In the case of the collared peccary, home ranges (in Arizona) of between one-fifth of a square mile and one-and-a-half square miles have been observed.[16] Around water holes in arid regions and near preferred bedding sites, ranges may overlap to the extent of between 100 and 200 meters, but in general herds occupy discrete territories.

The purpose of the peccary's scent gland is even now not fully understood. That it serves to demarcate territory is the most widely supported view.[17] But it has also been suggested that the scent plays some part in courtship,[18] that it "co-ordinates the herd's movements,"[19] and that it acts as an "alarm signal."[20] Apparently both species have a keen sense of smell. Some early commentators came to the conclusion that the orifice was part of the peccary's breathing system.[21] However, it was quite accurately, albeit briefly, described by Francisco Hernández (1571–1576),[22] and Montero de Miranda (1574) recognized the *ombligo* as a kind of scent gland (*respiradero hediondo*).[23]

Social organization. Studies of social organization (interaction, hierarchy, dominance, leadership) have been hindered by the difficulty of observing herds under natural conditions for a sufficient length of time. Among collared peccary, "a well-defined hierarchy can be observed in penned herds, but [this] is hard to detect in wild herds. Females are usually dominant over males."[24] Again, "strong following tendencies are apparent,

---

[15] "Largely nocturnal" according to Chapman, 1929: p. 212, 1931: p. 350; "generally nocturnal" in Ross, 1978: p. 8. Cf. Enders, 1935: p. 478 ("not nocturnal").

[16] Schweinsberg, 1971: p. 455; Sowls, 1974: pp. 144, 147–148, 154. Cf. Walker, 1975: 2: p. 1366 (range of 5 kilometers); Ellisor and Harwell, 1969: p. 426 (311 acres, 548 acres); Leopold (1959), 1972: p. 494 (some ranges "very circumscribed").

[17] Sowls, 1969: p. 220; Perry, 1970: p. 39; Schweinsberg and Sowls, 1972: p. 143; Sowls, 1974: pp. 144, 147–148, 154.

[18] Crandall, 1964: p. 528.

[19] Walker, 1975: 2: p. 1366.

[20] W. J. Hamilton, 1939: p. 88; Neal, 1959: p. 185.

[21] Léry (1557), 1592: p. 180, 1600: p. 137; León Pinelo (1650), 1943: pp. 51–52.

[22] Hernández (1571–1576), 1959: 1: pp. 310–311 ("un ombligo . . . y en el cual se junta un humor acuoso que fluye si se aprieta con los dedos").

[23] Montero de Miranda (1574), 1953: p. 348. Cf. Tyson, 1683: p. 378 ("scent gland"); Sánchez Labrador (ca. 1766), 1910: 1: p. 195 ("una prominencia en parte carnosa y en parte membranosa, por cual se transpira un humor tan fuerte olor de almizcle"); Herrera (ca. 1600), 1934–1956: 9: p. 335.

[24] Sowls, 1974: pp. 146, 149. Cf. Sowls, 1969: p. 220; Schweinsberg and Sowls, 1972: p. 132.

but . . . no clear evidence of leadership within herds."[25] It has been reported that when (? collared) peccaries migrate, younger animals are to the fore and the older bring up the rear.[26] White-lipped peccaries (typically in large droves) do appear to recognize a leader, usually an old animal, either boar or sow.[27]

[25] Sowls, 1969: p. 220, 1974: pp. 146, 162; Walker, 1964: 2: p. 1366. Nelson (1916: p. 448), however, refers to droves of collared peccary "usually led by the oldest and most powerful boar."

[26] Apparently the collared peccary in Hunn, 1977: p. 227; the white-lipped peccary in Leopold (1959), 1972: p. 488 (quoting Alvarez de Toro, 1952).

[27] Anon. (1607), 1908: p. 151; Herrera (ca. 1600), 1934–1956: 9: pp. 335–336 (Vera Paz, 1531); Holmberg, 1950: p. 25; C. A. Hill, 1966: p. 6; Córdova-Rios and Lamb, 1972: p. 56; Cadogan, 1973: p. 98.

PART II: THE PECCARY IN HUMAN ECONOMY AND SOCIETY

A: THE PRE-COLUMBIAN PERIOD

Osteological evidence

The bones and/or teeth of peccaries have been identified at several archaeological sites (map 4). Dates extend from the Preclassic (or Formative) to the period immediately preceding Spanish contact.[1]

Species	Period	Site	Region	Authority
	Preclassic	La Perra cave	Tamaulipas, Mexico	MacNeish, 1958: p. 140
? white-lipped	Preclassic	La Victoria	Guatemala	Coe, 1961: p. 141
collared	Preclassic	Parita Bay	Chiriquí, Panama	Willey, McGinnsey, 1954: p. 151
	Preclassic	Loma Alta	Guayas, Ecuador	Byrd, 1976: pp. 65, 108, 111
collared	Preclassic and Classic	Barton Ramie	Belize	Willey, Bullard, Glass, Gifford, 1965: pp. 523–524
collared	Preclassic and Classic	Dzibilchaltun	Yucatán, Mexico	Wing & Steadman, 1980: pp. 326–327
white-lipped	Classic	Dzibilchaltun	Yucatán, Mexico	Wing & Steadman, 1980: pp. 326–327
collared	Classic	Lubaantun	Belize	Wing, 1975: pp. 379–381
white-lipped and collared	Classic	Seibal	Guatemala	Pohl, 1976: pp. 97 ff.; Olsen, 1978: p. 174
white-lipped	Classic	Uaxactun	Guatemala	Ricketson & Ricketson, 1937: p. 204
collared and white-lipped	Classic	Uaxactun	Guatemala	Kidder, 1947: p. 60
collared and white-lipped	Classic and Postclassic	Altar de Sacrificios	Guatemala	Olsen, 1972: p. 244; Pohl, 1976: pp. 97 ff.
white-lipped	Classic and Postclassic	Kaminaljuyú	Guatemala	Kidder, Jennings, Schook, 1946: p. 157
white-lipped	? Postclassic	Coclé	Panama	Lothrop, 1937: p. 16

---

[1] "Remains of both peccaries are abundant in most Maya sites" (Olsen, 1982: p. 8). According to A. J. Ranere (1980: p. 31), collared and white-lipped peccaries were "probably available" to the inhabitants of pre-ceramic shelters in the Sierra de Talamanca, Costa Rica. General statements in Restrepo Tirado, 1892b: p. 110 (Colombia); Rojas González, 1949: p. 71 (Zapotec: Oaxaca, Mexico); García, 1952–1953: p. 158 (Totonac: Vera Cruz, Mexico); Olsen, 1964: pp. 23–24 (southern United States); Borhegyi, 1965: pp. 6, 23 (Guatemalan highlands); Bowen, 1976: p. 20 (Seri: Sonora, Mexico); Wing, 1977: p. 58 (Middle America and Trinidad).

Species	Period	Site	Region	Authority
collared	to Postclassic	Cozumel	Yucatán, Mexico	Hamblin, 1980; pp. 13, 223–247; 1984: pp 122–138
collared and white-lipped	Postclassic	Macanche	Guatemala	Pohl, 1976: fig. 4/3
collared and white-lipped	Postclassic	Mayapan	Yucatán, Mexico	Pollock, Roys, 1957: p. 639; Pollock, Roys, Proskouriakov, 1962: p. 377
collared and white-lipped	Postclassic	Tikal	Guatemala	Pohl, 1976: fig. 4/3
collared, ? and white-lipped		Cerro Brujo	Panama	Linares and White, 1980: pp. 182, 183, 186, 188

The animals were not necessarily obtained in the immediate vicinity of the above sites. In particular, the remains of white-lipped peccary at Dzibilchaltun and Mayapan suggest the possibility of live specimens or of carcasses brought in from a distance.[2]

The relatively high percentage of immature specimens at Cozumel (Yucatán) and at Cerro Brujo (Panama)[3] point perhaps to the capture and raising of juveniles, for which there are contemporary parallels in both areas. In the opinion of M. E. D. Pohl, "the high concentration on peccaries in the Postclassic Period at Flores could signify that these animals were tamed for eating or ritual purposes."[3a] At Seibal a walled structure may have been used as an animal pen.[4] "Rough stone troughs," recovered from several Mayan sites, could have served to water a variety of tame or penned animals, including peccary.[5]

Iconographic evidence

A rare model of a peccary (fig. 3) was found in the Preclassic of Tlatilco, Valley of Mexico,[6] and "a head resembling a wild pig" in the Preclassic of Seibal, Petén.[7] Pedro Simón (1626) refers to a ceremonial

---

[2] Pollack and Ray, 1957: p. 639. Analysis of faunal debris from archaeological sites on Cozumel (Hamblin, 1980: p. 223, 1984: pp. 123, 126) has not produced evidence of white-lipped peccary.

[3] Hamblin, 1980: pp. 237, 320, 1984: p. 132; Linares and White, 1980: p. 186.

[3a] Pohl, 1976: p. 205.

[4] Pohl and Feldman, 1982: p. 299; see also Pohl, 1976: pp. 5, 196, 203 ("penned peccaries" in Mayan folklore); Hamblin, 1980: p. 240, 1984: p. 123. Pollack and Ray (1957: p. 640) remark that "peccary remains [at Mayapan] seem particularly frequent in association with dwellings of the aristocracy."

[5] Gann, 1918: p. 55.

[6] Preclassic figures of animals, including peccaries, are mentioned by Piña Chan, 1971: p. 174.

[7] Willey in Willey (ed.), 1978: p. 12 (Escoba-Real phase).

MAP 4. Pre-Columbian sites with osteological evidence of peccaries.

FIG. 3. Pottery model of a peccary, Preclassic, Tlatilco, Valley
of Mexico. Photograph courtesy
of the Museum of the American
Indian, Heye Foundation.

site (*templo*) in the territory of the Cipacua (northern Colombia) "en que
adoraban un puerco espín de oro fino" (? peccary rather than porcupine).[8]
A realistic representation of a peccary appears in a band of glyphs over
a doorway on the eastern façade of the Casa de Monjas at Chichen Itza,
Yucatán.[9] Also at Chichen Itza there is a carving of a head-dress in the
form of a peccary,[10] and at Copan, Honduras, that of a monster with
the body of a man and the head of a peccary.[11]

Peccaries are portrayed in three pre-hispanic codices, two Mayan
(Dresden and Tro-Cortesianus) and one Mixtec (Nuttall).[12] They are
identifiable "by their prominent snout, curly tail, bristling dorsal crest,
and rather formidable tusks, as well as by the possession of hoofs,"[13]
but the two species cannot be distinguished. In Dresden there are two
peccary-head glyphs,[14] a peccary seated on a serpent,[15] and a creature
that combines the hoofs and bristles of the peccary and the scales of a
reptile.[16] Three panels show the peccary in association with the sky.[17]
Nuttall includes a strikingly realistic representation (fig. 4), as well as a

[8] Simón, 1882–1892: 1: p. 367. G. Hernández de Alba (1948b: p. 337) gives "peccary."
[9] Maudslay, 1889–1892: 3: pl. 13.
[10] Maudslay, 1889–1892: 3: pl. 45. See Tozzer and Allen (1910: p. 353) on "stone mask-
like figures" that may represent peccaries, in façade decoration in northern Yucatán.
[11] Maudslay, 1889–1892: 1: pl. 46. The peccary does not appear in G. Kubler's list of
animals in *The Iconography of the Art of Teotihuacan* (1967: p. 14). In a study of the Conte
style from Panama, O. F. Linares (1976a: p. 11) concluded that "the general rule seems to
be that animals that were eaten were not used in iconography."
[12] For studies of animal representation, see Stempell, 1908; Seler, 1909; Tozzer and Allen,
1910.
[13] Tozzer and Allen, 1910: p. 351.
[14] Dresden (Förstemann), 1880: pp. 43b, 45b; Thompson, 1972: p. 43b.
[15] Dresden (Förstemann), 1880: p. 62; Stempell, 1908: p. 712 [fig. 7]; Seler, 1909: p. 403
(fig. 240); Tozzer and Allen, 1910: pl. 33 (6).
[16] Tozzer and Allen, 1910: pl. 32 (6).
[17] Dresden (Förstemann), 1880: pp. 44b, 45b, 68a; Tozzer and Allen, 1910: p. 353, pls.
32 (2), 32 (4); Thompson, 1972: pp. 44b, 45b.

FIG. 4. Peccary in the *Codex Nuttall* (Mixtec, Pre-hispanic), Z. Nuttall, 1902: p. 73, col. 2.

peccary with human hands and feet.[18] The most arresting of the illustrations in Tro-Cortesianus are of peccaries caught in a noose (fig. 5).[19] It is possible that such animals were taken alive.

In the post-hispanic Codex Mendoza (1535–1550), the name-glyph for Ixcoyamec or Ixcoyametl is made up of a conventional eye (*ixtli*) superimposed on the body of a peccary (*coyámetl*).[20] The only other known references to the peccary in native sources are in the *Popol Vuh*

---

[18] Nuttall, 1902: pp. 73, 9 respectively; Seler, 1909: p. 402 [fig. 239].
[19] Tro-Cortesianus (Anders), 1967: pp. 49a (Stempell [1908: p. 711] suggests agouti), 49c, 93c (peccary also on 30b and possibly 66); Seler, 1909: p. 404 (figs. 244a, 244b); Kelley, 1976: p. 120.
[20] Codex Mendoza (Cooper Clark), 1938: 1: p. 8, 2: p. 54 (no. 595), 3: no. 595. Cf. Carrión (1581, ed. J. García Payón), 1965: p. 99.

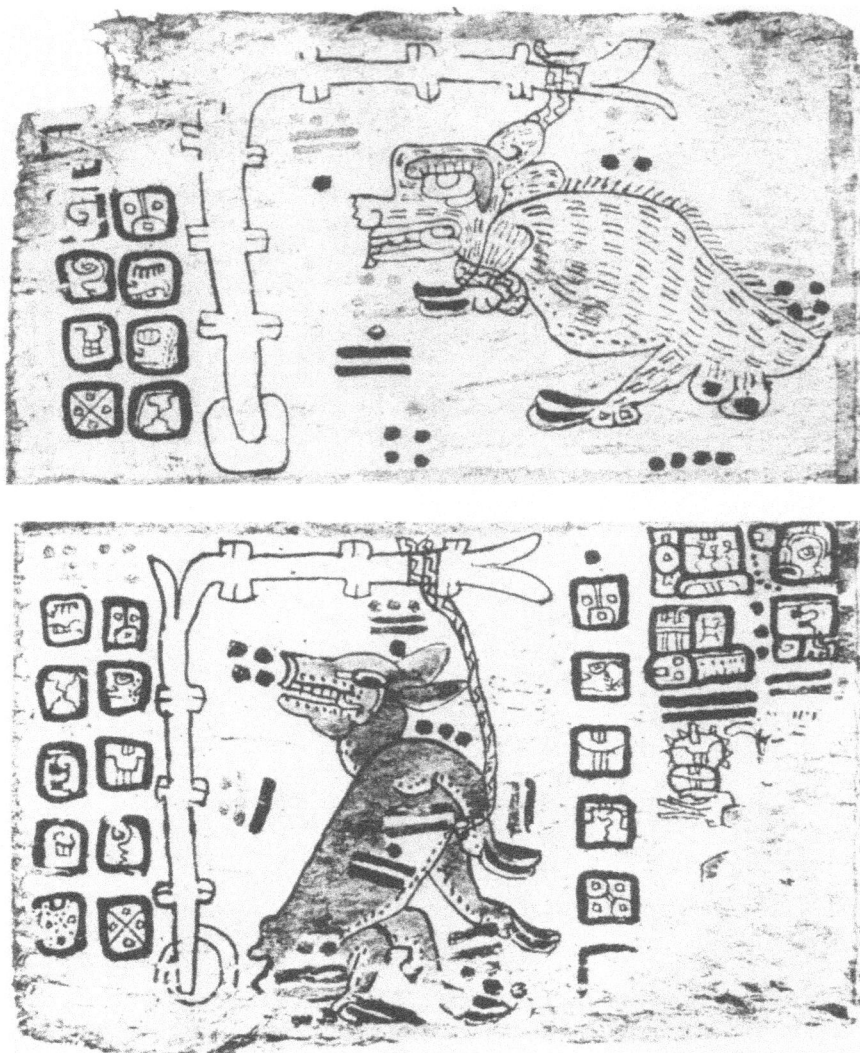

Fig. 5. Peccaries caught in a snare. *Codex Tro-Cortesianus* [*Madrid*] (Mayan, Pre-hispanic), E. Förstemann, 1902: pp. 49a, 49c.

of the Quiché of Guatemala[21] and in the *Book of Chilam Balam of Chumayal*.[22]

---

[21] Jena, 1944: p. 190 (*nim ac*); Edmonson (ed.), 1971: pp. 4, 21 (*nim aq*, "great pig" [white-lipped peccary]). The *Popol Vuh* is of pre-hispanic origin; the text used by Edmonson dates from 1550–1555.

[22] Roys, 1933: pp. 96, 130 (ca. 1490–1510, and successive compilations up to the end of the eighteenth century).

B: EUROPEAN CONTACT

Early reports of peccaries

The peccary is mentioned in many sixteenth- and seventeenth-century sources (map 5), most of the natural histories and a substantial number of the *relaciones geográficas*.[1] Before 1600 very few accounts make it clear that there were two species,[2] but thereafter reports from the humid tropics (based either on personal observations or on hearsay evidence and inferences drawn from the native nomenclature) often refer to both.[3] Generally the amount of supplementary information provided is very small, chiefly statements to the effect that the peccary was recognizably different from the wild pig of the Old World (notwithstanding some possible confusion with feral pigs), that they were commonly hunted, and that they, in turn, raided cultivated fields and gardens.[4]

There are, however, a number of important exceptions to these unremarkable observations. Reports from the coastlands of the southern

---

[1] Anon., 1900: pp. 36, 106; Paso y Troncoso (ed.), 1905a: pp. 66, 76, 237, 241, 246, 251, 1905b: pp. 64, 92, 104, 107, 111, 150, 181, 199, 1979: pp. 62, 121, 190, 320; Latorre y Setién (ed.), 1919: p. 86, 1920: pp. 50, 85, 93; Arellano Moreno (ed.), 1964: pp. 131, 169, 209; García Payon (ed.), 1965: p. 34; Jiménez de la Espada (ed.), 1965: 1 (1): p. 77, 2 (2): pp. 37, 80, 2 (3): pp. 213, 239, 3 (3): p. 61, 3 (4): pp. 199, 246, 277.

[2] Staden, 1557: 2: cap. XXX; Cardim (1590–1600), 1906: pp. 450–451, 1925: p. 37; Soares de Souza (1587), 1945: pp. 136–138; Montero de Miranda (1574), 1953: p. 348.

[3] Laet (1633), 1640: p. 484 (Brazil); Biet (1652), 1664: p. 340 (Cayenne); Acuña (1639), 1698: p. 69 (Amazon); W. (M.) (ca. 1699), 1732: p. 297 (Miskito Coast, Nicaragua); Warren (1667), 1752: p. 925 (Surinam); Dampier (1681), 1906: 1: p. 41 (Miskito Coast, Nicaragua); Anon. (1607), 1908: p. 151 (Panama); Harcourt (1613), 1928: p. 95 (Guiana); Wafer (1680–1688), 1934: pp. 15, 64, 102–103 (Panama); León Pinelo (ca. 1650), 1943: pp. 50–51 (Tierra Firme); Cobo (1653), 1956: 1: pp. 358, 363–364; Abbeuille (1614), 1963: p. 249 (Ilha Maranhão); Jiménez de la Espada (ed.), 1965: 3 (4): p. 246 (1619, Maynas, eastern Ecuador/Peru); Ruiz Blanco (ca. 1690), 1965: p. 23 (Piritú, Venezuela).

[4] Regional observations in: Thévet, 1558: p. 95, 1568: p. 77a, 1575: 2: p. 936b (Cayenne); Lerio (1557/1578), 1592: p. 180, 1600: p. 137 (Brazil and Nicaragua); Vargas Machuca, 1599: p. 153; Fleckmore, 1654: p. 71 (Brazil); Coreal (1666–1697), 1722: 1: p. 196 (Brazil); Ralegh (1595), 1848: p. 111 (Guiana); Cieza de León (1532–1550), 1853: p. 400, 1864: p. 174 (Portoviejo, Ecuador); Núñez Cabeza de Vaca [Pero Hernández] (1541–1544), 1891: pp. 118, 135, 1906: 1: pp. 182, 205 (La Plata); U. Schmidt (1535–1552), 1891: pp. 15, 19 (La Plata/Paraná); López de Velasco (1571–1574), 1894: pp. 431 (Quito), 442 (Piura, Peru), 454 (Santiago de las Montañas, Peru), 507 (Santa Cruz de la Sierra, Bolivia), 554 (La Plata), 557 (Ciudad Real [del Guaira], Paraguay), 565 (Brazil); Castillo (1675), 1906: p. 305 (Mojos); Baltasar de Ocampo (1610), 1907: p. 235 (Mañaries, Vilcapampa, Peru); Vázquez de Espinosa (ca. 1628), 1942: pp. 61 (Guiana), 208 (Chiapas), 644 (Santa Cruz de la Sierra, Bolivia); López de Gómara (1551–1552), 1954: 1: p. 154 (La Plata); Toribio de Ortiguera (1580–1590), 1968: p. 302 (Río Marañon, Peru).

MAP 5. Locations of references to peccaries in sixteenth- and seventeenth-century sources. *Relaciones geográficas* (1–33) in accompanying list.

Relaciones geográficas
1. Santiago de León [de Caracas] (1572/1578)
2. Trujillo (1579)
3. Nueva Zamora (1579)
4. Santa Cruz de la Sierra (1571)
5. Quixos [Quijo] (1570s)
6. Piura (16th century)
7. Asunción (1594)
8. Quito (1573)
9. Otavalo (1582)
10. [Río] Coraguana (16th century)
11. Los Maynas (1619)
12. Los Mojos (1564)
13. Vera Cruz (ca. 1571)
14. Valladolid (1570s), Sucopo (1579), Popola (1569)
15. Hueytlalpan (1581)
16. Tucumán (16th century)
17. Chinantla (1579)
18. Mitla[n]tongo (1579)
19. Puerto de Guatulco (16th century)
20. Pochutla (16th century)
21. Tonameca (16th century)
22. Guatulco (16th century)
23. Chila (1581)
24. Coyatitlanapa [in Aguatlan] (1581)
25. Xalapa de Vera Cruz (1580)
26. Xilotepec (1580)
27. Chepultepec (1580)
28. Tetela (1581)
29. Chilapa (1582)
30. Coatepec de Guerrero (1579)
31. Coatepec-Chalco (1579)
32. Huexutla de Hidalgo (1580)
33. Zumpango del Río (1582)

Caribbean indicate that peccaries were kept, fattened, and traded. In what is probably the earliest reference to the peccary, we learn that Columbus, off the coast of Costa Rica (*Cariay*) in 1503, took aboard two specimens "which an Irish wolfhound had not the courage to molest."[5] According to Fernández de Oviedo y Valdés (1514–), writing of Tierra Firme and more particularly the Isthmian province of *Cueva* (which he knew personally), "sucking pigs" (*lechones*) were occasionally captured.[6] But the most remarkable testimony comes from the lands around the Gulf of Urabá. In the vicinity of San Sebastián de Buenavista there were Indian traders, *grandes mercadores y contratantes*, who exchanged fish and salt and native pigs ("muchos puercos de los que se crian en la misma tierra") for the gold and cotton cloth of tribes further inland.[7] It is very doubtful whether the "pigs" were actually bred in captivity; more likely, our informant, Pedro de Cieza de León (ca. 1532), implies that captured juveniles were reared and fattened. A contemporary, Pascual de Andagoya, observed that in a particular valley on the Pacific coast (territory of the Barbacoa, Colombia) each dwelling had its pen for peccaries ("corrales de puercos de los naturales de allá").[8]

Similar statements are made by Antonio de Herrera (ca. 1600)[9] and Pedro Simón (1623–1626). Simón is particularly informative. We are told that the Urabá fattened *puercos de monte* in their houses ("engordan en sus casas"),[10] and that the Guazuzú were "requíssimas de oro fundido, que lo habían en rescates [exchange] de . . . puercos de monte cebados y gordos."[11] The Abibe, too, appear to have been involved in the traffic

[5] Columbus, 1825–1829: 1: pp. 284 [in Diego de Porras's account], 307 (1503), 1870: pp. 200–201, 1930: p. 302, 1963: pp. 353 [in Ferdinand Columbus's account], 381 (1503). Amerigo Vespucci (1893: p. 18) refers to "pigs" (*porci*) in the account (written 1504) of his alleged first voyage to the New World in 1497–1498 (landfall south of Trinidad and Paria). Other animals mentioned he could not have seen. The record is essentially imaginary (Markham in Vespucci, 1894: pp. i–xliv). However, he appears to have incorporated some information gathered on his "second" (in fact, first) voyage with Alonso de Hojeda (estuary of the Orinoco and Gulf of Paria) in 1499–1500, and it is possible that he then sighted or heard of peccaries.

[6] Fernández de Oviedo y Valdés, 1950: p. 152, 1959a: pp. 50–51, 1959b: 2: pp. 45–46. Andagoya ([1514–1546], 1865: pp. 17, 24), Espinosa ([1519], 1873: p. 32), Enriquez de Guzman ([1534–1535], 1862: p. 89), and López de Velasco ([1571–1574], 1894: p. 153) also refer to peccaries in Panama (Cueva/Coiba).

[7] Cieza de León (1532–1550), 1853: p. 361, 1864: p. 37. Fernández de Enciso (*Suma de Geografía* [1519], 1948: pp. 221–222) mentions gold (Darién) and salt (Isla Fuerte) and "muchos puercos" (Darién and Santa Marta), but not trade in these products. See also Benzoni (1541–1556), 1572: p. 79, 1857: p. 115 (Cueva and Darién); Barlow (1521–1531), 1932: p. 175 (Darién); López de Gómara (1551), 1954: 1: pp. 96, 118 (Darién); Las Casas (1559), 1951: 2: p. 410 (Urabá).

[8] Andagoya, 1829, p. 449 (opposite the island of Gallo, located on map 16). J. E. S. Linné (1929: pp. 130–131) suggested that peccaries were kept as "domestic animals" on the Pearl Islands off the Pacific coast of Panama. For this, however, there is no supporting evidence.

[9] Herrera, 1934–1956: 10: p. 102 (San Sebastián de Buenavista, 1532), following Cieza.

[10] Simón, 1882–1892: 5: p. 172.

[11] Simón, 1882–1892: 5: p. 173.

in gold, cotton *mantas*, and *puercos zahínos* between the coast and the hinterland.[12] Immediately to the east, in the middle Cauca valley, the Yamicí "raised captured peccaries in their houses."[13]

A description (1610) of the shorelands of the Laguna de Chiriqui (Panama), south of Columbus's landfall at Cariay, mentions trade in tame (*mansos*) peccaries and tapirs and other commodities (*caraña* and *chaquira*), again for inland gold.[14] The animals were valued for feasts and celebrations (*convites y fiestas*), particularly the much larger and rarer tapir.[15] Commerce involving reared peccaries may have been commonplace in lower Central America and northernmost South America, as far east as *Curiana*.[16] This is a region for which there are many later references to tame animals and birds, as well as to the Muscovy duck, the only species domesticated in the humid tropics of the Americas.[17]

Francisco Hernández (1571–1576) observed that the *coyámetl* of New Spain was tameable ("una vez que se domestica es apacible, se aficiona a los de casa y se granjea su cariño"),[18] and in Tlaxcala (1531), according to Herrera, "many peccaries" were bought and sold.[19] While there is no reason to doubt this, no similar observation for any other part of Middle America has come to light.

---

[12] Simón, 1882–1892: 4: p. 84. Núñez de Balboa, 1513 (trans. C. R. Markham in Andagoya, 1865: p. xv [Davaive]) mentions the exchange of fish for maize.

[13] Simón, 1882–1892: 5: p. 77.

[14] Zevallos, 1886: p. 157 (Río Tariric [Tarire, Sixaola] to the Escudo de Veragua). *Caraña* was a resinous, aromatic gum; *chaquira*, "mock pearl" or a kind of glass bead. The nature and significance of pre-Columbian trade in Panama and northern Colombia are considered by Helms, 1979: pp. 38–69, especially pp. 47, 56, 66. See also Linares, 1977: p. 73.

[15] Cf. Fernández Guardia, 1913: p. 21 (among the principal articles of commerce in pre-Columbian Costa Rica were "tapirs and wild hogs, domesticated for killing at their festivals"). López de Velasco ([1571–1574], 1894: pp. 328, 330) mentions peccaries in Costa Rica.

[16] See Caulin (1779), 1966: 1: pp. 73 ("y cogido alguno, se amansa, y domestica como los puercos caseros, à quienes se agregan, y muestran sociables"), 76, 78. Early references to peccaries in Anghiera (1511–1530), 1944: pp. 82, 83, 298 (hunting in *Curiana*), and Fernández de Oviedo y Valdés (1520–1555), 1959b: 3: pp. 80, 130, 164 (Cartagena, Santa Marta, Nuevo Reino de Granada). Kirchhoff (1948d: p. 483) observed that, north of the Orinoco, "people raised many animals in captivity. When young mammals refused to eat, women would feed them at their breasts." Thévet ([1558] 1568: p. 77) referred to a captured peccary in his account of Cayenne.

[17] Donkin n.d. (forthcoming).

[18] Hernández, 1959: 1: p. 311. Peccaries were first reported (1518–1521) from Cozumel (Yucatán) and the coasts of Tabasco/southern Vera Cruz (Torquemada [1615], 1723/1969: 1: p. 352; Anon. [1520s], 1858–1866: 1: p. 570; Leonardo de Argensola [ca. 1600], 1940: p. 47; López de Gómara [1551–1552]: 2: pp. 28, 54; Díaz del Castillo [ca. 1568], 1955: 1: p. 62). For Middle America, see: Gemelli Carreri (ca. 1700), 1704: 1: p. 548, 1955: 2: p. 207; López de Cogolludo (1688), 1867–1868: 1: p. 279 (Yucatán); Pomar (1582), 1891: p. 67 (Texcoco); López de Velasco (1571–1574), 1894: pp. 212 (Vera Cruz), 252 (Cozumel); Champlain (1599–1607), 1922–1936: 1: p. 58; Ponce (1588), 1932: p. 308 (Yucatán); Landa (ca. 1566), 1941: pp. 5, 204, 1966: p. 136 (Yucatán); Lázaro de Arregui (1621), 1946: pp. 42–44 (Nueva Galicia); Cervantes de Salazar (ca. 1560), 1971: 1: p. 126.

[19] Herrera, 1934–1956: 9: pp. 247–248 ("traen también a vender muchos puercos monteses, de los que tienen el ombligo al lomo, . . ."). The hunting of peccaries in Tlaxcala is mentioned by Muñoz Camargo (ca. 1576), 1892: p. 137.

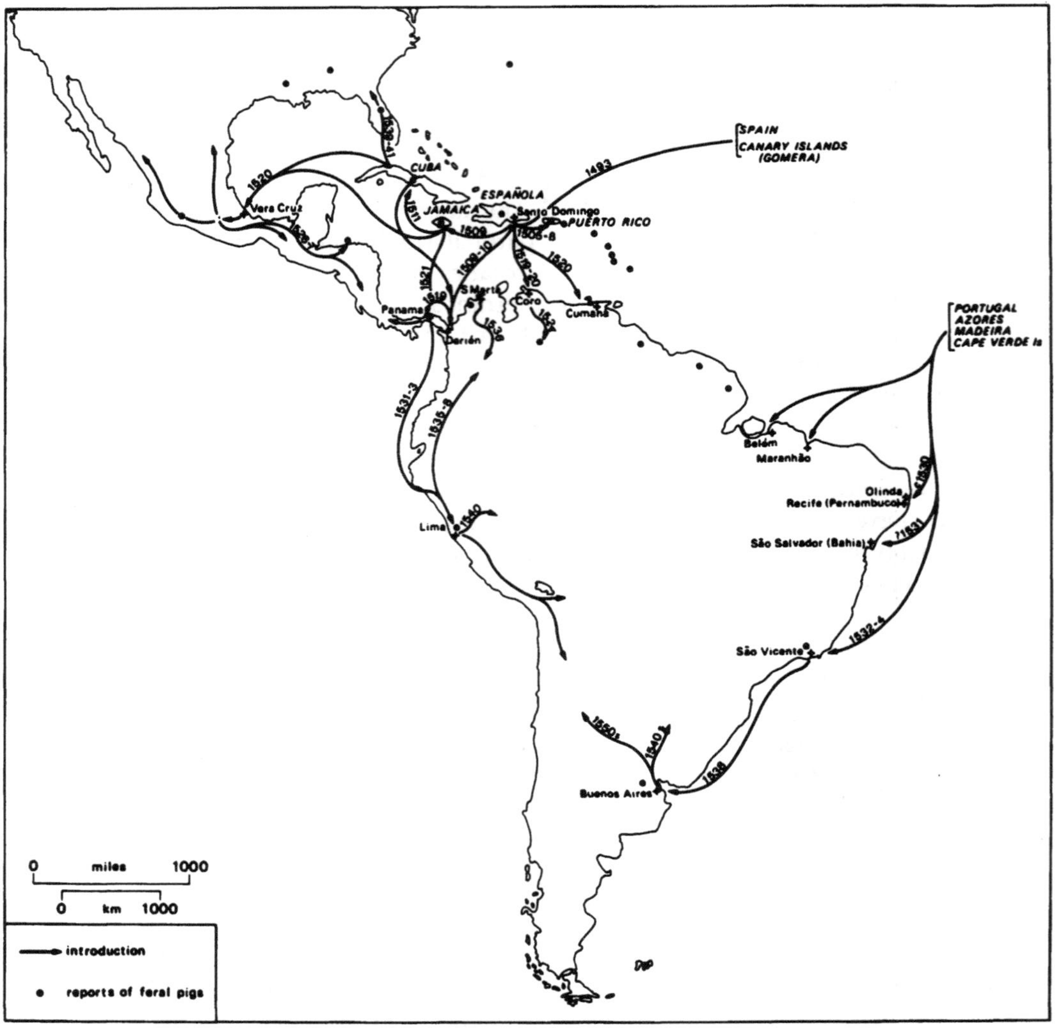

MAP 6. Pigs (*Sus scrofa domestica*) in the New World.

Pigs in the New World (map 6)

Pigs (*Sus scrofa domestica*) were among the first domesticated animals to be taken to the New World, and for several decades they were the most important species for food supply. The rearing of pigs had been part of the pastoral economy of southern Iberia since prehistoric times. Decline during the centuries of Moorish occupation was followed by renewed interest in the wake of the *Reconquista*.[20] From the hinterland

[20] Parsons, 1962: pp. 211–235.

of the ports of embarkation pigs were carried to the earliest island outposts of Spain and Portugal: to the Canaries, the Azores, Madeira, and the Cape Verdes. Livestock from the Atlantic colonies made up a significant proportion of what later passed to the Indies.[21]

Pigs (particularly the black- and russet-colored Iberian breeds, reared in open woodland) possessed pioneer qualities admirably suited to an age of exploration and colonization. They were omnivorous and, like the *conquistadores* themselves, hardy, mobile, and physically adaptable. They could more easily be transported by ship than other domestic mammals,[22] and later be driven overland to supply expeditions. They were also economical, in the sense that a high proportion of their body weight consisted of edible meat and lard.

Introduction to the New World: Columbus, on his second voyage to the Indies (1493), took on board eight pigs as well as other animals at La Gomera (Canaries),[23] all destined for the bridgehead in Española. According to Michele de Cuneo (28 October 1495) they "reproduced in a superlative manner, especially the pigs."[24] Other breeding stock followed,[25] and Spain's later mainland territories were supplied chiefly, if not exclusively, from populations established in the Islands. Swine were taken to Puerto Rico between 1505 and 1508,[26] to Jamaica from 1509,[27] and to Cuba (from Jamaica and possibly Española) in 1511.[28] During the following decade animals from one or more of these islands were dispatched to the southern shores of the Caribbean, from Darién[29] and Panama in the west to Coro[30] and Cumaná in the east. Pigs reached Panama first from Darién (1519) and then from Jamaica.[31] A royal grant to Panama in 1521 included 1,000 pigs from the *hacienda real* in Jamaica.[32] A decade or so later the Isthmian population provided the

---

[21] Wernicke, 1938: p. 78.

[22] See D. Navarrete (1671): 2: 1962: p. 356 (sucking pigs on board ship).

[23] Las Casas (1559), 1957: 1: p. 246 ("a setenta maravedís [of minute value, about one-quarter of a cent] la pieza").

[24] In Columbus, 1963: p. 217.

[25] Morrisey, 1957: p. 24; Deffontaines, 1957: p. 11. Las Casas clearly exaggerated in affirming "Destas ocho puercas se han multiplicado todos los puercos que hasta hoy ha habido y hay hoy en todas estas Indias. . . ."

[26] Coll y Toste, 1947: p. 62 (on the orders of Vicente Yañez Pinzón). The occupation of the island commenced in 1508 under Hernán Ponce de León.

[27] Sauer, 1966: p. 181 (by Juan de Esquivel and Francisco de Garay). See Morales Padrón, 1952: p. 276.

[28] Armas, 1888: p. 175; Sauer, 1966: p. 181 (by Diego de Velázquez de Cuéllar and Pánfilo de Narváez).

[29] Deffontaines, 1957: p. 11 (1509, presumably from Española). This date appears to be somewhat too early. Santa María la Antigua de Darién was "founded" in 1510 following the evacuation of the unsuccessful settlement (1509) of San Sebastián on the eastern side of the Gulf of Urabá. According to J. Ignacio de Armas (1888: p. 176) pigs were brought from Cuba (after 1511).

[30] Deffontaines, 1957: p. 11.

[31] Puente y Olea, 1900: pp. 438–439.

[32] Morrisey, 1957: p. 25 (quoting Cappa, 1889–1897: 5: p. 13, but not found at this reference).

initial breeding stock for Peru.[33] In 1540 Gonzalo Pizarro is said (by Garcilaso de la Vega) to have assembled 4,000 head of stock, "pigs and Peruvian sheep" (llamas), for an expedition to *La Canela*, "the land of Cinnamon," to the east of Quito.[34] Pigs may have been landed on the coast of what later became Colombia as early as 1516.[35] They apparently reached the central highlands between Popayán and Bogotá in the late 1530s from Peru and Ecuador (under Sebastián de Belalcázar)[36] and from the Caribbean coast (under Gonzalo Jiménez de Quesada).

For the conquest and settlement of Mexico swine were brought from Jamaica and Cuba and imported through Vera Cruz.[37] Cortés's expedition (1526–1527) to Honduras included a large herd of swine ("una gran manada de puercos"), some of which were released on the Bay Islands.[38] Likewise, Hernándo de Soto introduced pigs (from Cuba) to Florida ca. 1540.[39] Before about 1518 the authorities in Española attempted to maintain a monopoly of animal breeding, with the notable exception of pigs,[40] which multiplied very rapidly and, in the form of feral droves, ravaged cultivated crops, notably maize and sugar cane.

From the early years of the exploration of the New World the Spaniards and particularly the Portuguese released (*soltura*, from *soltar* "to turn loose") pigs and occasionally goats on uninhabited islands and stretches of inhospitable coast,[41] primarily for the benefit of mariners who might later be shipwrecked. Consequently the first settlers in such places (English and French, as well as Spaniards and Portuguese) often found feral livestock. Humphrey Gilbert (1583) credited the Portuguese with the introduction of swine to the island of Sablon or Sable (Nova Scotia).[42] Domingo Martínez de Irala (ca. 1540) left a sow and a boar on the island of San Gabriel in the estuary of the Río de la Plata and gave instructions that others should be placed on Martín García.[43] Some at least of the pigs introduced by Gonzalo de Mendoza (1538) to the province of La Plata appear to have come from Portuguese settlements

---

[33] Diego de Trujillo (1571), 1948: p. 50; Cobo (1653), 1956: 1: p. 285. Cobo has pertinent comments on the sale of pork (but little or no other meat) in the years following the foundation of Lima (1535).

[34] Garcilaso, 1966: 2: p. 874.

[35] Deffontaines, 1957: p. 11 (no authority).

[36] Roulin, 1835: p. 323. Roulin's paper has been largely forgotten, but is still useful.

[37] Armas, 1888: p. 176; Icaza, 1923: 1: p. 223 [Gregorio de Villalobos]; Hackett (ed.), 1923–1947: 1: 41; Dusenberry, 1963: p. 29; Chevalier, 1963: p. 85; Matesanz, 1965: pp. 535–536; Sauer, 1966: p. 189.

[38] Díaz del Castillo (ca. 1568), 1908–1916: 5: pp. 4, 16, 61–62, 1955: 2: p. 190; Conzemius, 1928: p. 62.

[39] Rye (ed.; 1557), 1851: pp. 85 [1541], 147; Hodge and Lewis (eds.), 1907: pp. 171, 235; Robertson, 1927: p. 12.

[40] Wernicke, 1938: p. 79.

[41] Tertre, 1667–1671: 2: p. 291; Roberts, 1726: pp. 418, 423; Barbot, 1732: p. 641 (quoting Antonio de Herrera, ca. 1601); Armas, 1888: p. 176; Wernicke, 1938: p. 79; Deffontaines, 1957: p. 9; Masefield, 1967: p. 280; Halley (1698–1701), 1981: pp. 45, 185.

[42] Hakluyt, 1904: 8: p. 63.

[43] Martínez de Irala, 1906: pp. 367–368.

on the coast of Brazil,[44] most likely São Vicente. In the course of the 1530s the Portuguese began landing livestock (first pigs and fowl, later other animals) from Olinda southward.[45] By about the middle of the century the descendants of pigs brought to La Plata had moved north to Paraguay and the region around Tucumán, near to the southern limits of expansion from Peru. Roulin's observation, in 1835, that, within half a century, domestic swine had spread to latitude 40 degrees south (Río Negro) and to latitude 25 degrees north (therefore omitting Florida) was substantially correct.[46]

Increase in numbers and the growth of feral populations. The important part that pigs played in the occupation of the New World was doubtless appreciated at the time,[47] and indeed evidence to this effect exists in various *relaciones.* Pork-butchers' shops (*carnicerias*) were quickly organized in Mexico City (1524)[48] and Lima (1536).[49] Pork was recommended, and might be reserved, for those who were ill, infirm, or convalescent.[50] Prices of live pigs—sometimes fed on tribute maize[51]—and of pig meat were at first high;[52] there was even a strong market in unborn sucking-pigs.[53] Then, as the number of animals increased, prices fell sharply,[54] sometimes to the point where it was hardly worth maintaining special pig farms (*hatos de puercos*[55]).

There is ample eyewitness support for José de Acosta's (1590) general observation that "swine . . . greatly multiplied in the Indies."[56] In Peru "sows proved very prolific."[57] Garcilaso de la Vega (1558) "saw two in the great square of Cuzco with thirty-two sucking pigs, each having put down sixteen."[58] Around Popayán and Pasto (Colombia), in the early seventeenth century, there were "countless hogs, with which they supply this country and ordinarily export much to Lima, a distance of 400

---

[44] Wernicke, 1938: p. 81.
[45] Wernicke, 1938: p. 80; Deffontaines, 1957: p. 11.
[46] Roulin, 1835: p. 324.
[47] Pereyra (1928), 1966: pp. 122–123; Prado, 1967: pp. 231–232 (Brazil).
[48] Bejarano (ed.), 1877: pp. 5, 58 ff.
[49] Cobo (1653), 1956: 1: p. 385; Romero, 1959: p. 99.
[50] Cobo (1653), 1956: 1: p. 386; Armas, 1888: p. 176. This has also been reported from West Africa, where pigs were introduced by the Portuguese (De Bry, 1604: p. 81; cf. Marees, 1605: p. 61).
[51] Acosta (1590), 1880: 1: p. 283; Chevalier, 1963: pp. 85–86; Matesenz, 1965: p. 536.
[52] Cieza de León (1532–1550), 1852–1853: p. 377; Garcilaso (1609), 1966: 1: p. 585; Armas, 1888: p. 176.
[53] Cieza de León (1532–1550), 1852–1853: p. 377 (fetching 100 pesos or more in the region of the Río Magdalena). Garcilaso ([1609], 1966: 1: p. 585) remarked on Cieza's statement.
[54] Ibid.; Chevalier, 1963: pp. 85–86; Matesenz, 1965: p. 536.
[55] González Dávila (1518), 1864: p. 342 (Española). Ranches were at first known as *hatos,* and this is probably the earliest reference to pig farms in the Americas.
[56] Acosta, 1880: 1: p. 283; similarly Cobo (1653), 1956: 1: p. 385.
[57] Garcilaso (1609), 1966: 1: p. 585. J. J. von Tschudi ([1838–1842], 1847: p. 250, 1844–1846: 1: p. 255) reported feral pigs (*chanchos simarones*) in the vicinity of plantations around Lima.
[58] Garcilaso (1609), 1966: 1: pp. 585–586.

leagues."[59] The growth in numbers was also remarked in parts of Guatemala and Mexico.[60] By about 1700 more than 30,000 pigs were consumed annually in the city of Mexico.[61] An obstacle to the expansion of swine is mentioned by Father Ignaz Pfefferkorn (1794–1795), writing of Sonora. There no one would consent to be a swineherd.

To expect a Spaniard to become one would be a sovereign offense. And no Indian can be induced to do it, not because his pride stands in the way, but because of his inherent, implacable hatred for swine. The animal is so abhorrent to him that he would suffer the severest hunger rather than eat a piece of domestic pork.[62]

How widespread and persistent was the rejection of pork by the indigenous population of the New World has not been determined.

High fertility and feral populations were apparent within a decade or so in the restricted environments of the islands. In 1503 Nicolás de Ovando purchased the right to hunt *cerdos silvestres* in Española.[63] Oviedo (1526) reported that "many of the swine carried from Spain . . . [had] become wild . . . in Santo Domingo, Cuba, San Juan [Puerto Rico] and Jamaica."[64] At the same time, those "that escaped to the forests on Tierra Firme did not live long, for they were eaten by jaguars, ocelots and cougars." This presumably refers to that part of the mainland known to Oviedo personally (*Cueva*, the Caribbean littoral).

The situation in Española was confirmed by Las Casas (ca. 1550), by Acosta (ca. 1590) and by Cobo (1653).[65] Edward Topsell (1607) repeated the opinion that the swine there grew "to the stature of mules."[66] In the mountains of Jamaica there were "countless herds . . . fair game to anyone who chose to kill them."[67] Diego de Velázquez (1514) claimed that the swine that he had introduced to Cuba had increased within three years to the barely credible total of 30,000.[68] In Puerto Rico, too,

---

[59] Vázquez de Espinosa (ca. 1628), 1942: p. 357 (no. 1081).

[60] Fuentes y Guzmán (1690), 1969–1972: 2: p. 242; Cortés y Larraz (ca. 1769), 1958: 2: p. 83; Ajofrín (ca. 1765), 1958–1959: 2: p. 172: Clavijero (ca. 1770), 1968: p. 495. Vázquez de Espinosa ([ca. 1628], 1942: p. 177) referred to "native wild swine" (peccaries or feral pigs ?) in Michoacán.

[61] Vetancurt (1698), 1960–1961: 2: p. 194 (17).

[62] Pfefferkorn, 1949: p. 103.

[63] Armas, 1888: p. 175. Sauer (1966: p. 157) has 1508.

[64] Fernández de Oviedo y Valdés (*Sumario*, 1526), 1950: pp. 151–152, 1959a: pp. 50–51. Cf. ibid. (*Historia General y Natural*, 1520–1555), 1959b: 2: p. 184 (Jamaica). In the *Sumario*, Oviedo describes feral pigs as "puercos monteses" (usually applied to peccaries). Similarly, in the *Historia General*, "puercos salvajes" refers to both feral pigs (Jamaica) and peccaries, or *báquiras* (1959b: 3: pp. 80 [province of Santa Marta], 164 [province of Cartagena]).

[65] Las Casas, 1958: 1: pp. 18, 20; Acosta, 1880: 1: p. 283, 1962: p. 306; Cobo, 1956: 1: p. 386.

[66] Topsell, 1607: p. 665. The record of feral droves continues to modern times (Gabb [1870], 1881: p. 125).

[67] Vázquez de Espinosa (ca. 1628), 1942: p. 117 (no. 332). López de Velasco ([1571–1574], 1894: p. 120) referred to "puercos de que hay muchos cimarrones." For later reports, see [Le Sieur] Thomas, 1674: p. 9; Sloane, 1707–1725: 1: p. xvi; Browne, 1789: p. 487.

[68] Armas, 1888: p. 175; Sauer, 1966: p. 189.

animals multiplied prodigiously, and feral pigs were reported in 1582.[69] On some of the smaller islands there were serious ecological and economic repercussions. In Bermuda (1594) the swine were "so lean that you cannot eat them, by reason the island is so barren,"[70] and a slightly later account (ante 1622) referred to damage to crops by feral pigs.[71] There are also reports, spread over several centuries but pointing in the same direction, for Barbados,[72] St. Thomas,[73] Dominica,[74] Sable,[75] the Falklands,[76] the Bay Islands,[77] and the French possessions of Martinique, Guadeloupe, and St. Christophe.[78]

Around the margins of eastern South America feral populations were less conspicuous, but nevertheless did not go unremarked.[79] Here there were also effective predators (as first observed by Oviedo) and the risk of confusion with peccaries. In the interior of the continent, the introduction and adoption of pigs were often long delayed.[80] Some groups still do not keep them and others treat the pig (and other domestic species) as a pet that is rejected as a source of food.

Pigs of African origin. Marcgravius's *Historiae Rerum Naturalium Brasiliae* appeared in Leiden in 1648. The author's patron was Joan Mauritz,[81] governor of the Dutch possessions in Brazil (1637–1644), and Marcgravius's direct knowledge of Brazil was largely, if not exclusively, confined to the region around Pernambuco (Recife). Immediately following the section devoted to the peccary (*tajacu*), Marcgravius described and illustrated (fig. 6) *Porcus Guineensis,*[82] the Guinea or Red river-hog, *Potamochoerus porcus* (one of three genera of African suids). This, we are told, had been introduced ("translatus") to Brazil from West Africa as a

---

[69] Latorre (ed.), 1919: p. 48. Cf. Coll y Toste, 1947: pp. 63–64.

[70] May, 1904: p. 202.

[71] Anon. [? John Smith], 1822: pp. 13, 230. See also Gates (1609–1610), 1906: pp. 22–23 ("thousands of wilde hogges"); Lefroy (1621), 1877–1879: 2: p. 583.

[72] Crosby, 1972: p. 78; but cf. Ligon, 1657: p. 59.

[73] Moseley (1872–1876), 1879: p. 324.

[74] Atwood, 1791: p. 47 (by then "hunted down by negroes"). D. Taylor (1938: p. 149) refers to "wild pigs," presumably feral.

[75] Hakluyt (Gilbert, 1583), 1904: 8: p. 63.

[76] Garnot, 1826: p. 41.

[77] Díaz del Castillo (ca. 1568), 1908–1916: 5: pp. 61–62 (ca. 1530).

[78] Tertre, 1667–1671: 2: pp. 291–298; Labat (1725–1727), 1730: 3: p. 312; Roulin, 1835: pp. 328–329.

[79] Bancroft, 1769: p. 124 (Guiana); Labat (1725–1727), 1730: 3: p. 312 (Cayénne); Quandt, 1807: pp. 201–202 (Surinam); Rengger, 1830: p. 331 n. (province of Buenos Aires); Southey, 1810–1819: 3: p. 852 (? "wild boars," São Paulo); Roulin, 1835: pp. 324–325, Paez, 1868: p. 143 (*llanos* of Venezuela). For southeastern North America (few predators and no peccaries), see Gosse, 1859: pp. 63, 270 (Alabama); Dureau de la Malle, 1855: pp. 806–807 (Louisiana); Parsons, 1962: p. 228 n. (Florida).

[80] J. B. Turner, 1967: p. 138 (to the Northern Kayapó in 1879, quoting Coudreau, 1897).

[81] Count of Nassau-Siegen (1604–1679), sometimes known as Maurice of Nassau; not to be confused with Maurice (1567–1625), son of William the Silent.

[82] Marcgravius, 1648: p. 230 (long tail, pointed ears, "russi coloris"). Marcgravius was born in Liebstadt, Saxony (1610), and died in Guinea (1644).

PORCVS GVINEENSIS, & è Guinea in Braſiliam tranſlatus,figura ut noſtrates & ruffi coloris : in hoc autem differ tà noſtratibus, quod caput habeat non ita elatum: aures autem longas

& acutas plane & prolongatis acuminibus, caudam longam uſque ad talos propendentem. pi-
lorum expertem. Totum corpus tegitur pilis brevibus ruffis ſplendentibus, non ſetis, quibus
& in dorſo caret, ſed tantum verſus caudam in dorſo & circa collum paulo longiores habet pi-
los. Plane cicur.

FIG. 6. The Guinea or Red river-hog, *Potamochoerus porcus*. Georgius Marcgravius,
1648: p. 230.

tame ("plane cicur"), but not explicitly domestic, animal.[83] It is possible
(as suggested by J. Reinhardt[84]) that Marcgravius only saw the African
hog in the menagerie maintained by Mauritz at his residence in Pernambuco. In any event, the Guinea hog is not mentioned by any earlier
authority, such as Gabriel Soares de Souza (1587),[85] or indeed by
Marcgravius's contemporary and collaborator Gulielmus Piso.[86] Further-
more, it is not entirely clear that later statements were based on
information other than that provided by Marcgravius.

"The American Hog," wrote John Hill (1752),

is a native of many parts of South America; it is not only wild in the woods, but
is kept tame about houses for the sake of its flesh. . . . The writers on the
Brazilian animals have described it, and from them others. Ray and most of the
moderns have called it *Porcus Guiniensis*, the Guinea Hog.[87]

Patrick Browne (1756), the Comte de Buffon (1767), and J. V. P. Erxleven
(1777) each affirmed that the Guinea hog was numerous or prolific in
the Americas.[88] "Prolific" cannot apply to the peccary, but could be said

---

[83] Commenting on Marcgravius, J. E. Gray (1868: pp. 36–37) has "naturalized," F. J.
Simoons (1953: p. 80) "entirely domesticated."

[84] Reinhardt, 1869: pp. 56–57.

[85] *Notícia do Brasil*, 1587 (1945: 2 vols.).

[86] *Indiae Utriusque Re Naturali et Medica*, Amstelaedami, 1658.

[87] J. Hill, 1752: p. 572. The first part of the statement suggests possible confusion with
the peccary, which, however, is separately described under the "Musk Hog." *Porcus
Guineensis* Marcgr. in Ray, 1693: p. 96.

[88] Erxleven, 1777: p. 185 ("Brasiliam translatus, ubi hodie copiosisimus cicur"); Browne
(1756), 1789: p. 487 ("breeds a greater number of pigs than any other kind"); Buffon,
1884: 9: p. 235 ("multiplié en Amérique").

of both domestic and feral varieties of *Sus scrofa*, including any of Asiatic origin.[89] According to J. E. Gray, the Guinea hog will not breed with the domestic pig,[90] but there are also statements to the contrary.[91]

As well as being introduced as a curiosity, the Guinea hog (and likewise the Guinea fowl and chicken with black flesh) may have reached the New World as deck cargo intended as provisions;[92] the same has been suggested for the Muscovy duck, moving in the opposite direction.[93] The American duck and the Guinea fowl were, of course, domesticated; the Guinea hog, like the peccary, is occasionally tamed. Tame specimens of *Potamochoerus porcus* have been reported from the northeast Congo (among the Mangbetu, the Abarambo, and the Niam Niam),[94] the lower Niger, and Liberia.[95] Whether or not associated with the slave trade, African pigs could have been transported to South America at any time after the middle decades of the sixteenth century. At present, however, the importance of their contribution to the suid population, domestic or feral, remains obscure.

---

[89] P. L. S. Müller [Linnaeus], 1773–1776: 1: p. 465. Humboldt ([1811] 1966: 3: p. 51) states that pigs were introduced from the Philippines as well as from Europe. According to Cappa (1889–1897: 5: pp. 428–429), "De China se trasladaron los puercos y perros que allí se crian; los puercos son menores. . . . No parece que se ha perpetuado esta casta." Sanderson (1950: p. 781) refers to Chinese pigs (? feral) in Surinam. The question of Asiatic introductions invites further investigation.

[90] Gray, 1868: pp. 36–37.

[91] Laurence, 1805: p. 510; Law, 1842: 2: p. 18; Youatt, 1860: p. 176; Johnston, 1905: p. 2, 1906: 2: p. 720. According to H. von Nathusius (1864: pp. 171–172), English agricultural writers from the middle of the eighteenth century remarked that the Red Pig of Guinea was used for cross-breeding.

[92] Cappa, 1889–1897: 5: p. 428.

[93] Donkin n.d. (forthcoming).

[94] Schweinfurth, 1873: 2: p. 89; Hartmann, 1884: p. 185; Czekanowski, 1917–1924: 2: p. 219; Johnston, 1908: 2: p. 616. See also Malbrant [French Equatorial Africa], 1952: p. 40; Boettger, 1958: pp. 105, 139, 278. Domestic pigs (*Sus*) are relatively common in non-Islamic areas of West Africa (Joubert and Bonsma, 1961: p. 155, Abb. 73).

[95] Johnston, 1905: p. 202, 1906: 2: pp. 718–720, 1908: 1: p. 192 n. 4; Schwab, 1947: p. 69. The distribution of wild *P. porcus* extends south to the lower Congo (Zaire) and Angola. Another species, *P. koiropotamus* (the "Bush-pig"), belongs to the scrublands of East and South Africa.

C: FOLK NOMENCLATURE

Luso-Hispanic names (map 7)

The most common European names for the peccary derive from its physical resemblance to the domestic pig (Spanish *puerco, cerdo;* Portuguese *porco*) and the wild boar (Arabic/Spanish *jabalí,* hence *javelin, javelina;* Portuguese *javalí*). They date from the earliest knowledge of the animal in the territories around the southern and western shores of the Caribbean (Tierra Firme). *Jabalí* was employed by Pietro Martire d'Anghiera (ca. 1510) in accounts of Coiba (Panama) and Curiana (Venezuela).[1] The name may refer to either or both species of peccary, but properly only to the larger and more gregarious, made more explicit by the description *jabalí de manada* (troop or drove). The Anglicized form, in use along the Caribbean coast of Central America and in Guiana, is *warree (warí, warrí).*[2]

Puerco de monte generally refers to the collared peccary, *puerco de manda (manada),*[3] *puerco de tropa* and *puerco jabalí* to the white-lipped peccary. *Puerco de la tierra*[4] presumably includes both. Portuguese *porco do matto* is usually the collared peccary, but a distinction may be drawn between the *porquinho do matto* (collared) and the *porco do matto* (white-lipped).[5] The latter is more commonly known as *porco de queixo branco* ("white-jawed pig") or simply *queixada,* referring either to its distinctive jaw or to its habit of clashing the teeth. The comparable Spanish description, *cariblanco (carrillo blanco,* "white cheeked"), is apparently confined to Costa Rica.[6]

---

[1] Anghiera [first coll. ed. of the *Decades,* 1516; based on interviews, not personal observations], 1944: pp. 82, 83, 298. See also Acosta (1590), 1962: pp. 199–200.

[2] Bartholomew Sharpe (1680) in Burney, 1803–1807: 4: p. 94 (Isthmus); Dampier (1681), 1906: 1: p. 41 (Miskito coast, Nicaragua); Wafer (1680–1688), 1934: p. 64 (Isthmus); W. (ca. 1699), 1732: p. 297 (Miskito coast); Bancroft, 1769: p. 125 (Guiana); Stedman (1772–1777), 1796: 1: pp. 355–356 (Surinam); T. Young, 1842: p. 170 (Miskito coast); Squier (1855), 1969: p. 192 (? Honduras, Costa Rica, Nicaragua); Conzemius, 1932: p. 165 (Miskito coast); Frost, 1974: p. 159 (Tortuguero, Costa Rica). The *warree* is usually contrasted with the smaller "peccary."

[3] Sometimes abbreviated to *manado, manada* (Anon. [1607], 1908: p. 151, Panama; Le Roy Gordon, 1957: p. 72, Sinú, Colombia).

[4] López de Velasco (1571–1574), 1894: pp. 328, 442, 557; Anon., 1900: p. 106 [Yucatán, 1579]; Torquemada (1615), 1723: 1: p. 352, 2: p. 297.

[5] Tastevin, 1923: pp. 702, 740. Soares de Souza (1587; 1945: 2: pp. 136–138) appears to include both species under *porco do mato.*

[6] First noted in Frantzius, 1869: p. 296. Also the common name of a monkey, *Cebus hypoleucus,* in South America.

48

MAP 7. Hispanic regional names for the peccary.

Associations with colloquial names for "hog" include *chancho*, thus *chancho del monte*, the collared peccary,[7] but also the common name of the agouti, *Dasyprocta* spp., in Peru; *cochino* [*coche, cuche, cuchi*] *de*

[7] Perhaps the *chuche* of Fernández de Oviedo y Valdés (1526), 1950: p. 152 (1520–1555), 1959b: 2: p. 45 (Cueva, eastern Panama). See also Azara, 1802: 1: p. 25 (*chuncho* and *cosqui*, Paraguay); Paz Soldan, 1862: pp. 186 (*chancho*, Piura, Peru), 686 (*chancho del monte*, collared peccary, Loreto, Peru); Ambrosetti, 1894a: pp. 701–702 (*chancho jabalí*, Misiones, Argentina); Stone, 1949: p. 41 (Costa Rica). Cf. *chacharo* (collared peccary) of northwest South America, in Maximilian, 1825–1832: 2: p. 561 (the Ature and Maipure); Santamaría, 1942: 3: p. 453 (the Maipure); Barker, 1953: p. 443 (the Waica); Altolaguirre y Duvale (ed.), 1954: p. 126 (Valle de Aroa, 1768); Rohl, 1959: p. 139 (Venezuela); M.-C. Muller, 1975: p. 69 (Venezuela).

monte;[8] and *marrano* [*de monte*], thus *moran, marino, moro*.[9] In Nicaragua *jagüilla* ("little jaguar") is sometimes used to describe the white-lipped peccary.[10]

Two other Spanish names are rather more obscure. One is *tatabro* or *tatabaro*, peculiar to Ecuador and Colombia (? and Darién[11]). F. J. Santamaria distinguished between *tatabaro* (collared peccary in Ecuador) and *tatabro* (white-lipped peccary in Colombia).[12] The earliest known use of the latter (or indeed of either) is in Juan de Velasco's *Historia del Reino de Quito* (1789).[13] There appears to be some connection with *tayasú, taitetú* (Tupí-Guaraní, collared peccary),[14] and *tatú*, "armadillo."[15]

The name *saíno* (*sayno, sajone, sagino, sajino, saijino, saguino, zaíno, zainu, zahino, zayno, zajino, zagino, cayno, caino, cahino*) was in use from at least the middle of the sixteenth century. Cieza de León (ca. 1550) has *puercos zainos* (Gulf of Urabá, Colombia),[16] López de Velasco (1571–1574) *puercos caynos* (Caribbean coast of Panama).[17] In 1619 *puercos zaynos de monte* were reported from the province of Maynas in the eastern lowlands of Peru/Ecuador.[18] José de Acosta (1590) refers simply to *los saynos*[19] and this, or a cognate, was subsequently applied to one or other or both species of peccary throughout early Tierra Firme[20]—from southeastern Mexico[21] and Central America[22] to northern South

---

[8] Alston, 1879–1882: p. 107 (*coche de monte*, Guatemala); Simson, 1886: p. 264 (? *cáshi*, Záparo, eastern Ecuador); León, 1912: sect. xix (*cuchingâná*, Popoloca, Mexico); Lehmann, 1920: 2: p. 675 (*cuchi, cushe*, Lenca, Honduras); Abregú Virreira, 1942: p. 254 (*kita ccuchi*, Aymara); Velasco (1789), 1946: 1: pp. 118–119 (*ituchi, huasi-cuchi, pucuchi, casha-cuchi*); Appun (1871), 1961: pp. 189, 457 (*cochino de monte*); Dyk and Stoudt, 1965: p. 90 (*cuchí yúcú*, "wood hog," Mixtec, Mexico); Orr and Wrisley, 1965: p. 20 (*lumu cuchi*, Quichua, eastern Ecuador).

[9] Pineda, 1845: p. 24 (Chiapas); Alston, 1879–1882: p. 107 (collared peccary, Mexico); Gadow, 1908: p. 374 (? both species, southern Mexico).

[10] Armas, 1888: p. 68; Lehmann, 1920: 1: p. 112. Cf. Brigham (1887), 1965: p. 370 (? Guatemala).

[11] Cullen, 1868: pp. 161–162 (*tatabro*, collared peccary).

[12] Santamaría, 1942: 3: p. 142, 1959: p. 892 (*tatabro*, Colombia); Armas, 1888: p. 68 (white-lipped peccary, Colombia). S. A. Barrett (1925: 1: p. 12) also has *tatabara* (? collared peccary, Ecuador). *Tatabro* (collared peccary, Colombia) in Restrepo Tirado, 1892a: p. 5; Wassén, 1935: p. 86; West, 1957: p. 162 (? large droves); and Real Academia Española *Diccionario* (1970 ed.).

[13] Velasco, 1946: 1: p. 118 (white-lipped peccary). Not included in Boyd-Bowman *Lexico Hispanoamericano del siglo XVI* (1971).

[14] B. T. Solari (1928: p. 140) has *tateto* = *tatabra*.

[15] See infra p. 52 n. 33.

[16] Cieza de León, 1862: p. 361.

[17] López de Velasco, 1894: p. 353.

[18] Jiménez de la Espada (ed.), 1965: 3 (4): p. 246.

[19] Acosta, 1962: pp. 205–206.

[20] Simón (1626), 1882–1892: 1: p. 372 (*puercos zahínos*); León Pinelo (ca. 1650), 1943: 2: p. 51 (*zainos*); Cobo (1653), 1956: 1: p. 364 (*zahino*). Cf. Nieremberg, 1635: pp. 170–171; Johnstone, ca. 1650: tab. XLVI.

[21] Santamaría, 1959: pp. 892, 995. Probably more widely distributed. Pfefferkorn ([1794–1795], 1949: p. 112), writing of Sonora, gives "*sayno* or more commonly *javalie*." Also perhaps *senso*, the white-lipped peccary in Chiapas (Pineda, 1845: p. 24; Alvarez de Toro, 1952: p. 192; Leopold [1959], 1972: pp. 497, 500; Anon., 1976: p. 29).

[22] Frantzius, 1869: p. 269, Alston, 1879–1882: p. 107, Armas, 1888: p. 67, Wagner, 1958:

America[23]—where the greatest variety of Hispanic names occurs. Thence the distribution extends southward along the Andean *montaña* and adjacent lowlands,[24] perhaps as far as the Gran Chaco.[25] *Saino* may be derived from *saín* ("fat"), *sainar* ("to fatten"),[26] or from *zaino* ("dark-colored" and/or "vicious," of animals[27]), or from *cano* ("grey-haired"),[28] or again from *seno* ("cavity"),[29] referring to the peccary's prominent scent gland.

Vernacular names

Middle and South America gave birth to several hundred languages. Many can be grouped into major or minor families; others appear to be independent or are of doubtful affiliation. A substantial number have disappeared since they were discovered and partly recorded in the sixteenth century or later. Available vocabularies differ widely in provenance and quality and employ different systems of transliteration. Etymological studies are rare. Tracing the origins and connections of particular names is correspondingly difficult.

Where both species of peccary are present in a particular area, they are usually, if not invariably, given separate (but often related) names. A choice of more than two names[30] suggests distinctions based on superficial appearance, age, or sex; and/or the use of synonyms, including hybrid forms with Spanish or Portuguese or other Amerindian elements; and/or confusion with feral pigs (*Sus scrofa*).

South America

The most important language families from the point of view of peccary nomenclature are both South American: Tupí-Guaraní and

---

p. 242, Santamaría, 1959: p. 892 (Costa Rica); Anon. (1607), 1908: p. 151, Cullen, 1868: pp. 161–162, Goldman, 1920: p. 73, Allen and Barbour, 1923: p. 261, Enders, 1935: p. 469, Aldrich and Bole, 1937: p. 186, F. Johnson, 1948: p. 234 [1610], Bennett, 1968b: p. 40, Mendez, 1970: p. 243 (Panama, including Darién).

[23] Jeffreys, 1762: p. 17 (Cartagena); Reclus, 1881: p. 214, Armas, 1888: p. 67, Restrepo Tirado, 1892a: p. 5, 1892b: p. 110, Brettes, 1903: p. 340, Wassén, 1935: p. 86, Le Roy Gordon, 1957: p. 72, West, 1957: p. 162 (Colombia); Hernández de Alba, 1948d: p. 476, Ruíz Blanco (ca. 1690), 1965: p. 23 (North-central Venezuela); Kirchhoff, 1948d: p. 483 (Orinoco); Velasco (1789), 1946: 1: p. 118; Barrett, 1925: 1: p. 12 (Ecuador).

[24] Paz Soldan, 1862: p. 532 (*puercos sajinos*, Loreto, eastern Peru); Rice, 1910: p. 687 (*saina*, Río Vaupés); Juan Dueñas (1792) in Izaguirre Ispizua, 1922–1929: 8: p. 245 (*saginos*, eastern Peru); Jiménez de la Espada (ed.), 1965: 3 (4): p. 246 (*puercos zaynos de monte* [1619], eastern Peru); Pierret and Dourojeanni, 1967: p. 15 (*sajino*, eastern Peru); Denevan, 1972: p. 171 (*sajino*, Gran Pajonal, eastern Peru); Campos, 1977: p. 59 (*sajino*, Río Pisqui, eastern Peru); Roe, 1982: 98 (eastern Peru).

[25] Lozano, 1733: p. 40.

[26] Friederici, 1947: p. 671.

[27] Explicit in Anon. (1607), 1908: p. 151.

[28] Santamaría, 1959: p. 892 (quoting Herrera).

[29] Pfefferkorn (1794–1795), 1949: p. 112.

[30] C. W. Beebe (1917: p. 405), writing of the Uruata region of Guiana, mentions that the Indians name five "kinds" of peccary.

Cariban, contributing *tayasú* and *pecarí* respectively. These and their recognizable cognates are more widely distributed than other names. *Tayasú* and *pecarí* were incorporated in the early scientific nomenclature. They have also given rise to loan words in Portuguese (*caitetú*) and Spanish (*báquira*), which may complicate the problem of establishing what local names are employed.

Tupí-Guaraní (map 8)

Tayasú (the *Tayassu* [1814] and *tajacu* [1758] of science). This has been recorded from the very extensive Tupí-Guaraní realm over a period of more than 400 years. It is not surprising, therefore, that there are many variants. At the beginning of the nineteenth century Félix de Azara pointed out that, in Paraguay, *tayazú* alone may refer to either or both of the peccaries or to the domestic or feral pig.[31] Generally, however, it is reserved for the peccaries, more often than not the larger species (thus similar to Spanish *jabalí*). The elements are *tāi* (tooth) and *açu* [*guaçú*, *guasu*] (large).[32]

The earliest known reference to the name (*teygaju dattu*) is in Hans Staden's (1547–1555) account of eastern Brazil.[33] Jean de Léry (1557), also on the basis of first-hand knowledge, gives *taiassou*,[34] and Soares de Souza (1587) two species, *tajaçuété* or *tajaçuguita* (white-lipped) and *tajaçutirica* (collared).[35] From the second half of the seventeenth century, naturalists quoted chiefly Marcgravius, who had briefly described and quaintly illustrated *tajacu caaigoara*[36] ("of the woods or *monte*," *ka'aguy*), the collared peccary. Similarly *kuré* ("pig") may be combined with

---

[31] Azara (1801), 1838: pp. 115, 123. Cf. ibid., 1802: 1: pp. 24–25, 1809: 1: p. 248; Martius, 1863: p. 477 (*tayasu* v. *tayacu aya* = *Sus domesticus*, porco manso); Tastevin, 1923: p. 740; Solari, 1928: p. 150; Stradelli, 1928: p. 296 (*taiasuáia*, *taiasú-suaia*).

[32] B. Caetano (ed.) in Cardim, 1925: p. 112; Paraja da Silva (ed.) in Soares de Souza, 1945: 2: p. 137. Curiously, and apparently erroneously, Amuchástegu (1966: xvi) derives *tajasú* from *taya* (? *Descurainea canescens*), the root of which is relished by the peccary.

[33] Staden, 1557: part II, chapter XXX (no pagination); *teygasu tattu* in Staden, 1592: p. 129, *teygasu dattu* in Staden, 1874: p. 160, and *tanhaçú-tatú* in Staden, 1963: p. 148. The armadillo (*tatú*), subject of the following chapter, is called *dattu* in 1557, *tattu* in 1592, *dattu* in 1874, and *tatú* in 1963. Vázquez de Espinosa ([ca. 1628] 1942: p. 674) referred to "animals like pigs, which they call *tatuus* (armadillos)." Resemblances between the armadillo and a "sucking pig" were remarked by Pfefferkorn (Sonora, 1794–1795 [1949: p. 114]). The collared peccary sometimes takes refuge in the burrow of the giant armadillo.

[34] Lerio, 1592: p. 180, 1600 [first published 1578]: p. 137 (Brazil and ? Nicaragua). Followed by Laet, 1633: p. 551. *Tajassouh* in Thévet, 1575: 2: p. 936 b.

[35] Soares de Souza, 1945: 2: p. 136 (-*été*, superlative suffix). Cf. Laet, 1640: p. 484; Martius, 1867: 2: p. 87 (*taiaçu été*, *taiçuiété*); Ruiz de Montoya (1639), 1876 (*tayaçu eté*); Cardim (1590–1600), 1925: p. 37 (*tayaçutirica* and *tayaçupita*); Abregú Virreira, 1942: p. 254 (*tayacú-eté*); Abbeuille (1614), 1963: p. 249 (*tayássou été*). *Tirica* also appears as *tiragua* (Cabrera and Yepes, 1940: p. 280).

[36] Marcgravius, 1648: p. 229. *Caaigüara* according to Azara, 1802: 1: p. 25; *tayacú-uira* [*guira*] in Martius, 1867: 2: p. 478. See also Cabrera and Yepes, 1940: p. 280.

MAP 8. Tribal and linguistic distributions. A—Apiaca; C—Cainguá; Ca—Calianá; Ch—Chiriguano; Co—Cocama; M—Mundurucú; O—Omagua; Oy—Oyampí; S—Sirionó; T—Tapirapé.

ka'aguy.[37] Another qualifying word is *tinga*,[38] implying a pungent odor (of the white-lipped peccary).

Staden's *teygaju dattu* reappears as *tayasú-titu*[39] and, much more commonly, as *taitetú* or *caitetú*—the collared peccary. Seventeenth- and eighteenth-century sources, commencing with Ruiz de Montoya's *Vocabulario de la lengua Tupi ó Guaraní* (1639), refer to the former or one of the many cognates;[40] *caitetú* belongs to the nineteenth century.[41] The corresponding name for the white-lipped peccary, *tānicati* (*tañihca-tí*, *tañíg-catí*, *tagnicate*), is apparently first recorded by Azara (ca. 1800);[42] it combines the Guaraní words for "jaw" (*tanyka*) and "whiteness" (*tí*), the Portuguese *queixo branco* or *queixada*.

Modern authorities attest to variants of *tayasú* (or of *taitetú*, *caitetú*) in use among the Mundurucú,[43] the Omagua,[44] the Apiaca,[45] the Calianá,[46] the Cainguá,[47] the Sirionó,[48] the Kagwahiv,[49] the Tapirapé,[50] the Cocama,[51] the Oyampí,[52] and the Chiriguano.[53] Adoption by tribes speaking languages other than Tupí-Guaraní seems to be rare, although some sources are equivocal. According to A. Simson (1886), *tayasó* is the name for the white-lipped peccary among the Piojé,[54] a Tucanoan-speaking people in eastern Ecuador, to the northwest of the Omagua.[55]

---

[37] Azara (1801), 1838: p. 115 (*curé* alone); Solari, 1928: p. 110; Dennler, 1939: p. 140 (*curé-caá-bîg*); Abregú Virreira, 1942: p. 254 (*curé caagüi*); Guasch, 1961: p. 293 (*kure ka'aguy*).

[38] Martius, 1863: p. 481 (*tayasu-tinga*); Burton (ed.) in Staden, 1874: p. 160 n. (*tayu tinga*).

[39] Martius, 1863: p. 481, 1867: 2: p. 87, *taiatytú; tayasú taitetú* in Dennler, 1939: p. 239.

[40] See also Fleckmore, 1654: p. 71 (*tatoo*); Eckart, 1785: p. 512 (*taytetu*); Azara (1781–1801), 1802: 1: pp. 23–29, 1809: 1: p. 249, 1838: p. 122 (*taytetú*).

[41] Maximilian (1815–1817), 1820: p. 174, 1825–1832: 2: p. 588; Martius, 1863: pp. 477, 481; Liais, 1872: p. 404; Burton (ed.) in Staden, 1874: p. 160 (*caa-eté* "virgin forest" and *cuu*[*suu*] "game," "the 'c' being changed for euphony into 't' ").

[42] Azara, 1809: 1: p. 349, 1838: p. 119.

[43] Martius, 1863: p. 477, *tathié;* Coudreau, 1897b: p. 196, *iradié-tiou;* Strömer, 1932: p. 29, *dad'ekt'u, kaititu* (collared peccary), *dad'e, dad'e waụaụa* (white-lipped peccary); Nimuendajú, 1932: p. 99, *ẓadektyú* (collared peccary), *radyê* (white-lipped peccary); Murphy and Murphy, 1974: p. 63, *catitú* (collared peccary).

[44] Martius, 1863: p. 477 (*tajacú, tayasú*).

[45] Coudreau, 1897b: p. 187 (*tazaou*).

[46] Koch-Grünberg, 1916–1928: 4: p. 315 (*tẹhí*, collared peccary).

[47] F. Vogt, 1904: p. 209 (*tayačú*); Cadogan, 1973: p. 98 (Pâi-Cayuá, Chiripá, *tajasú;* Mbyá, *tajachú*).

[48] Holmberg, 1950: p. 24 (*čiásu*).

[49] Kracke, 1981: p. 137 (*tajahu, taitetu*).

[50] Wagley, 1977: p. 311 (*tachaho;* the *tanchahó* of Baldus, 1970: p. 212).

[51] Martius, 1867: 2: p. 300.

[52] Martius, 1867: 2: p. 323.

[53] Bayo, 1931: p. 189 (*tayasu*).

[54] Simson, 1886: p. 267.

[55] Cf. also Botocudo *curähk-niptiacu niómm*, white-lipped peccary (Martius, 1867: 2: p. 182).

MAP 9. Tribal and linguistic distributions. A—Apalai; Ar—Arára; B—Bacaïri; C—Cumanagoto; Ch—Chaima; G—Galabi; H—Hianácoto-Umáua [Carijona]; I—Ingarico; M—Macusi; Ma—Mape and Chaké; Mk—Makiritare; Mp—Mapoyo; P—Paraviyana; Pa—Panare; Pu—Puricoto; S—Sapará; T—Taulipang and Arecuna; Tr—Trio; W—Waiwai; Wa—Wayumara; Y—Yauarána [Yabarana].

Cariban (map 9)

The name "peccary" comes from Cariban *paki:ra*,[56] thus *báquira, pakir, pakira, patirá, pockiero, pecarí*, among the many variants. There may be an ancient connection with Tupí-Guaraní *kuré* (? *pag-kuré*).[57] Cariban *paki:ra*, unless used generically (as in the early literature), refers to the collared peccary; however, *báquira* or *váquira*, as a loan word in Spanish, is applied to the larger species.[58]

The name appears in very early accounts of Tierra Firme. First, there is an obscure passage in Columbus's record (7 July 1503) of his fourth voyage. Off the Caribbean coast of the Isthmus (Cariay-Veragua) he took on board a wounded monkey and a pair of "pigs" (clearly peccaries).[59] We are told that the former was known as *begare*. This is almost certainly a mistake,[60] either a syntactical error or a misunderstanding by Columbus and possibly others of the ship's company. If, however, *begare* corresponds to *paki:ra*, it is not immediately clear how it came to be current in this non-Carib area.[61] Fernández de Oviedo y Valdés (1520-1555), who knew the Isthmus well, says that the native pig of Cueva and Castilla del Oro (east of Veragua) was called *chuche* (? Spanish *cochino*), in "other provinces" of Tierra Firme *báquira*.[62] Elsewhere he singles out Santa Marta and Cartagena,[63] the nearer hinterland of which was not occupied by Caribs.[64] *Báquira* is also mentioned in a *relación* of the province of Caracas in 1572-1578.[65] Apparently the name had been widely diffused and adopted in the lands around the southern margins of the Caribbean.[66]

Evidence from the seventeenth and eighteenth centuries relates chiefly to the heartland of the Carib domain—Guiana,[67] Surinam,[68] Cayenne,[69]

---

[56] Hoff, 1968: p. 421.

[57] Dennler, 1939: p. 239. Cf. Liais, 1872: p. 401; Brand, 1964: p. 134.

[58] Simpson, 1941: pp. 14-15; Civrieux, 1980: p. 164.

[59] In Fernández de Navarrete (ed.), 1825-1837: 1: p. 307; Columbus (Italian ed.), 1864: pp. 333-334.

[60] Morison in Columbus, 1963: p. 381. *Begare* clearly refers to one or other animal. Nevertheless Armas (1888: p. 68) ventures another interpretation. *Echarle begare* (Italian *beccàre*), he maintained, referred to the act of throwing a dead or dying animal (in this case the monkey) to within reach of another animal (the peccaries); *dar beccàre* = *dar de comer*. It was, therefore, only "una coincidencia muy extraordinaria" that *begare* and *pecarí* are phonetically similar.

[61] Lehmann (1920: 1: p. 112) has *baquira* (?).

[62] Fernández de Oviedo y Valdés (1526), 1950: p. 152, (1520-1555), 1959b: 2: p. 45.

[63] Fernández de Oviedo y Valdés (1520-1555), 1959b: 3: pp. 80, 164.

[64] See Sauer, 1966: pp. 170-172 ("Caribs at Cartagena ?").

[65] Latorre (ed.), 1919: p. 86; Moreno and Arellano (eds.), 1964: p. 131.

[66] Vargas Machuca, 1599: p. 153 (*baquira*).

[67] Bancroft, 1769: p. 125 (*picary*); Harcourt (1608-1609/1613), 1928: p. 95 (*pockiero*); Vázquez de Espinosa (ca. 1628), 1942: p. 61 (*váquira*).

[68] Warren (1667), 1752: p. 925 (*pakeera*); Stedman (1772-1777), 1796: pp. 355-356 (peccary). Dampier (1681) (1906: 1: p. 41) also uses the word "peccary" (collared peccary) in writing of the Miskito coast of Nicaragua.

[69] Biet (1652), 1664: p. 340 (*paquira*); Labat [Desmarchais] (1725-1727), 1730: 3: p. 312 (*pekari*).

and neighboring parts of Brazil and Venezuela.[70] And from the same region and the same period come the earliest references to the complementary Carib name for the white-lipped peccary—*pingo* or *poingo*.[71] The distribution of Carib tribes known to employ, or to have employed, *paki:ra, pingo,* or some cognate name is similarly concentrated, except for the Bacaïri of central Brazil (? *pãhu; poséka, pohéka*):[72]

	collared peccary	white-lipped peccary	
Apalai (1924)	pakĭrä	poiñokŏ	73
Chaima (1680)	paquera	puuique	74
Cumanagoto (1680, 1683)	paquera, vaquira	puuique	75
Galibi (1867, 1971)	pockiero, pagi:la	pingo, pii:ngo	76
Ingarico (1928)	pakí:la	peyiṅgé, peyiṅkẹ́	77
Macusi (1847–1848, 1867, 1908, 1932)	peraka, paraka	poinké, puinká	78
? Mape and Chaké (1914)	báquiro		79
Mapoyo (1975)	pakira	poinke	80
Panare (1982)	paika	pinkë	81
Paraviyana (1847–1848, 1867, 1932)	pakira	poinké	82
Puricoto (1924, 1928)	pakilá	peyiṅgé, piyĭnkŭ	83
Sapará (1928)	pakí:la	peiṅgẹ	84
Taulipang and Arecuna (1928)	pakí:la	peyiṅgé, péiṅgo	85

---

[70] Tauste (1680), 1888: p. 36 (*paquera*); Yangues (1683), 1888: p. 157 (*vaquira*); Ruíz Blanco (ca. 1690), 1888: p. 175 (*vaquira*); Gumilla, 1791: 1: p. 259 (*paquira*); Cisneros, 1764: p. 22 (*baquira*); León Pinelo (1650), 1943: 2: p. 50 (*baquira*, provincias de Tierra Firme); Altolaguirre y Duvale (ed.) (1767–1768), 1954: pp. 117 [Barquisimeto], 126 [Aroa], 200 [Coro] (*baquira*).

[71] Biet (1652), 1664: p. 340 (*poinga*); Warren (1667), 1752: p. 925 (*pinko*); Cisneros, 1764: pp. 22, 26 (*pinque*); Fermin, 1765: 1: p. 12 (*pingo*); Stedman (1712–1717), 1796: pp. 355–356 (*pingo*); Tauste (1680), 1888: p. 36 (*puuique*); Harcourt (1608–1609), 1928: p. 95 (*paingo*); Altolaguirre y Duvale (ed.) (1768), 1954: p. 126 (*puinque*).

[72] Steinen, 1892: p. 37.

[73] Farabee, 1924: p. 232.

[74] Tauste (1680), 1888: p. 36.

[75] Tauste (1680), 1888: p. 36; Yangues (1683), 1888: p. 157; Ruíz Blanco (ca. 1690), 1888: p. 175.

[76] Martius, 1867: 2: pp. 362–363; Kloos, 1971: pp. 58–59.

[77] Koch-Grünberg, 1916–1928: 4: p. 252.

[78] Schomburgk (1840–1844), 1847–1848: 3: p. 784, 1923: 2: p. 130; Martius, 1867: 2: p. 226 (*pengŏu, paingŏu*); Koch-Grünberg and Hubner, 1908: p. 25; J. J. Williams, 1932: p. 212.

[79] Osgood, 1914b: p. 46.

[80] M.-C. Muller, 1975: p. 69.

[81] Henley, 1982: pp. 56–57. *Paika* and *pinkö* in Dumont, 1976: p. 57.

[82] Schomburgk (1840–1844), 1847–1848: 3: p. 784, 1923: 2: p. 130; Martius, 1867: 2: p. 228; J. J. Williams, 1932: p. 212.

[83] Farabee, 1924: p. 248; Koch-Grünberg, 1916–1928: 4: p. 264.

[84] Koch-Grünberg, 1916–1928: 4: p. 264.

[85] Koch-Grünberg, 1916–1928: 4: pp. 37, 252.

Trio (1972)	*pakira*	*ponjeke*	[86]
Waiwai (1963, 1965)	*pakrí*	*poinkó*	[87]
Yauarána [Yabarana] (1928)	*pákali*	*peínke*	[88]

Near neighbors of Cariban-speaking tribes who appear to have adopted *pecarí*-related names include the Warrau (*pakitye, pakilye*),[89] the Makú (*pékelya*),[90] the Maipure and Ature (*paquira*),[91] and the Arawakan Wapishana (*bakur*)[92] and Maopityan (*bakur*).[93] Jívaro *pákki* is the most isolated.[94]

Carib *haūÿa* ("jabali")[95] is represented in Azumara *hiya*,[96] Wayumará *hī:yé* (white-lipped peccary) and *haki:lá* (collared peccary),[97] Hianácoto Umáua *hakîta* (collared peccary);[98] and, among Arawakan tribes, in Manao *haya*,[99] Yavitéro *ahiya* (white-lipped peccary),[100] Mandauáca *ahida, ahī:da* (white-lipped peccary),[101] Juri *ahtä,*[102] and possibly Guinau *iuá:ra*.[103] *Haūÿa* may incorporate a root form of the word for peccary, in turn related to Arawak *abüya*.

Arawakan (map 10)

Arawak *abüya* (*abuia, abuie,* "to feed") is the white-lipped peccary.[104] The name is applied in the Guianas,[105] along the upper Orinoco,[106] and

---

[86] Lenselink, 1972: p. 40.

[87] Fock, 1963: p. 159; Yde, 1965: p. 122.

[88] Koch-Grünberg, 1916–1928: 4: p. 239.

[89] Schomburgk (1840–1844), 1847–1848: 3: p. 784.

[90] Koch-Grünberg, 1916–1928: 4: p. 321.

[91] Maximilian, 1825–1832: 2: p. 561.

[92] Farabee, 1918: p. 221.

[93] Farabee, 1918: p. 285. Possibly also Cayuishana *puicküé* (Martius, 1867: 2: p. 259) and Goajiro *pútschi, pána* (Jahn, 1927: p. 365).

[94] Karsten, 1920: pp. 9, 12, 1935: pp. 171, 570 (*pákki, yankípi,* collared peccary; *unta pákki,* white-lipped peccary); swine are known as *kuchi,* Quechua (in origin Spanish). The Jívaro occupy territory to the north of the Río Marañon, chiefly in southeast Ecuador.

[95] Lehmann, 1920: 1: p. 19.

[96] Farabee, 1924: p. 244 (unlocated).

[97] Koch-Grünberg, 1916–1928: 4: p. 264.

[98] Koch-Grünberg, 1908: p. 45.

[99] Martius, 1863: p. 477.

[100] Goeje, 1928: p. 226.

[101] Koch-Grünberg, 1916–1928: 4: p. 295.

[102] Martius, 1867: 2: p. 271.

[103] Koch-Grünberg, 1916–1928: 4: p. 283.

[104] Goeje, 1928: p. 226, *abüya* = "bush hog, taiasu, *Dicotyles labiatus*," the white-lipped peccary; however, in two of the Arawakan languages (Mariate, Marauha or Marawa) it is identified as *D. torquatus,* the collared peccary. Elsewhere, Goeje (1928: p. 259) has "bush hog, peccary, taitetú" [collared peccary] = *abüya* or *matúla.* Other authors also identify *abüya* (or derivatives) as collared peccary; perhaps the name has generic status.

[105] Schomburgk, 1847–1848: 3: p. 784, 1923: 2: p. 130 (*apuya*); Brett, 1851: p. 37 (*aboeya*); Thurn, 1883: p. 109 (*abouyah*); Beebe et al., 1917: p. 465 (*abuya*); Roth, 1924: pp. 182–183 (*abuya*). Cf. Maximilian, 1825–1832: 2: p. 561 (*apida*); Sargot, 1852: p. 67 (*aboia*).

[106] Note also Guahibo (non-Arawak) *ábutse* (Koch-Grünberg, 1916–1928: 4: p. 347), *gabuiza* (Chaffanjon, 1889: p. 322).

in the northwest of the basin of the Amazon. Variants of *abüya* have been recorded for the following members of the Arawak-Maipure linguistic family:

Adzáneni	*apija, apidza, ahida,*[107] *á:pidza*[108]
Arekena [Uarekena, Arekana]	*abida,*[107] *abîda*[109]
Baniva [Baniwa]	*abîda,*[109] *abida*[110]
Baré	*abida,*[107] *habîa, habîya,*[109] *abiatschy,*[111] *habija*[112]
Cariaya	*aputery*[113]
Carutana [Karutana]	*apitša,*[107] *âpitša*[109]
Catapolítani	*âpitsa*[109]
Cayuishana [Cauixana]	*putzya*[107]
Jumana	*apuya,*[107] *abúia*[114]
Maipure	*apia*[107]
Manao	*abiatschy*[113]
Marawa [Marauha]	*abia*[107]
Mariate	*apytza*[107]
Pasé [Passe]	*abaeghua*[107]
Piapóco	*apidza,*[107] *apitsa,*[115] *apítža*[116]
Siusi	*apitsa,*[107] *âpitsa, âpitse*[109]
Tariana	*apia,*[107] *âpia*[109]
Uainumá	*hapychtsche*[107]
Uirina	*abiaxe*[117]
Waraicú	*abûy*[118]

The white-lipped peccary is also known in the Guianas as *keherum, kairuni*.[119] Richard Schomburgk (1840–1844) derived this from the Macusi (Cariban) word *kair*, "evil smelling."[120]

On the upper Orinoco the collared peccary is called *chamu* (Achagua), *dzamulítu* (Adzáneni), *tzamúlitu* (Catapolítani), *dzamu* (Piapóco), *yamúlitu* (Tariana), and *samōlití* (Siusi and Carutana).[121] In the southwest Amazon

---

[107] Goeje, 1928: p. 226; Caulin (1779), 1966: 1: p. 73.

[108] Koch-Grünberg, 1916–1928: 4: p. 295.

[109] Koch-Grünberg, 1911: p. 132.

[110] Chaffanjon, 1889: p. 340.

[111] Martius, 1863: p. 477; Chaffanjon, 1889: p. 322.

[112] Martius, 1867: 2: p. 286.

[113] Martius, 1863: p. 477.

[114] Ibid.

[115] Chaffanjon, 1889: p. 328.

[116] Koch-Grünberg, 1916–1928: 4: p. 295 (also *apídža, apídza*).

[117] Martius, 1867: 2: p. 229.

[118] Martius, 1863: p. 477.

[119] Brett, 1851: p. 37 (*kaero*); Martius, 1867: 2: p. 311 (*keherum*); Thurn, 1883: p. 109 (*kairuni*); Beebe et al., 1917: p. 465 (*kairuni, karuata*). See also Schomburgk, 1837: p. 321 (*kairouni*); Roth, 1924: pp. 182–183, 479 (*kairuni, karuata*); Harris in Harcourt, 1928: p. 95n (*kairuni*); Perry, 1970: p. 39 (*kairui*).

[120] Schomburgk, 1923: 2: p. 74 (184); apparently followed by Roth, 1924: pp. 182–183. Cf. Schomburgk, 1847–1848: 3 [not translated by Roth, 1923]: p. 784 (*kairuni*, Arawak). Goeje (1928: p. 259) gives *kēerun*, "stinking thing?" (not associated with the Macusi).

[121] Goeje, 1928: p. 226; Koch-Grünberg, 1911: p. 132, 1916–1928: 4: p. 295. Also Gauhibo *tsámuli* (Koch-Grünberg, 1916–1928: 4: p. 347). Cf. among the southern and southwestern Arawak, Bauré *simori*, Caouiri *samouri*, Mojo *simoru*, Campa *samani* (Farabee, 1922: p. 49; Goeje, 1928: p. 226).

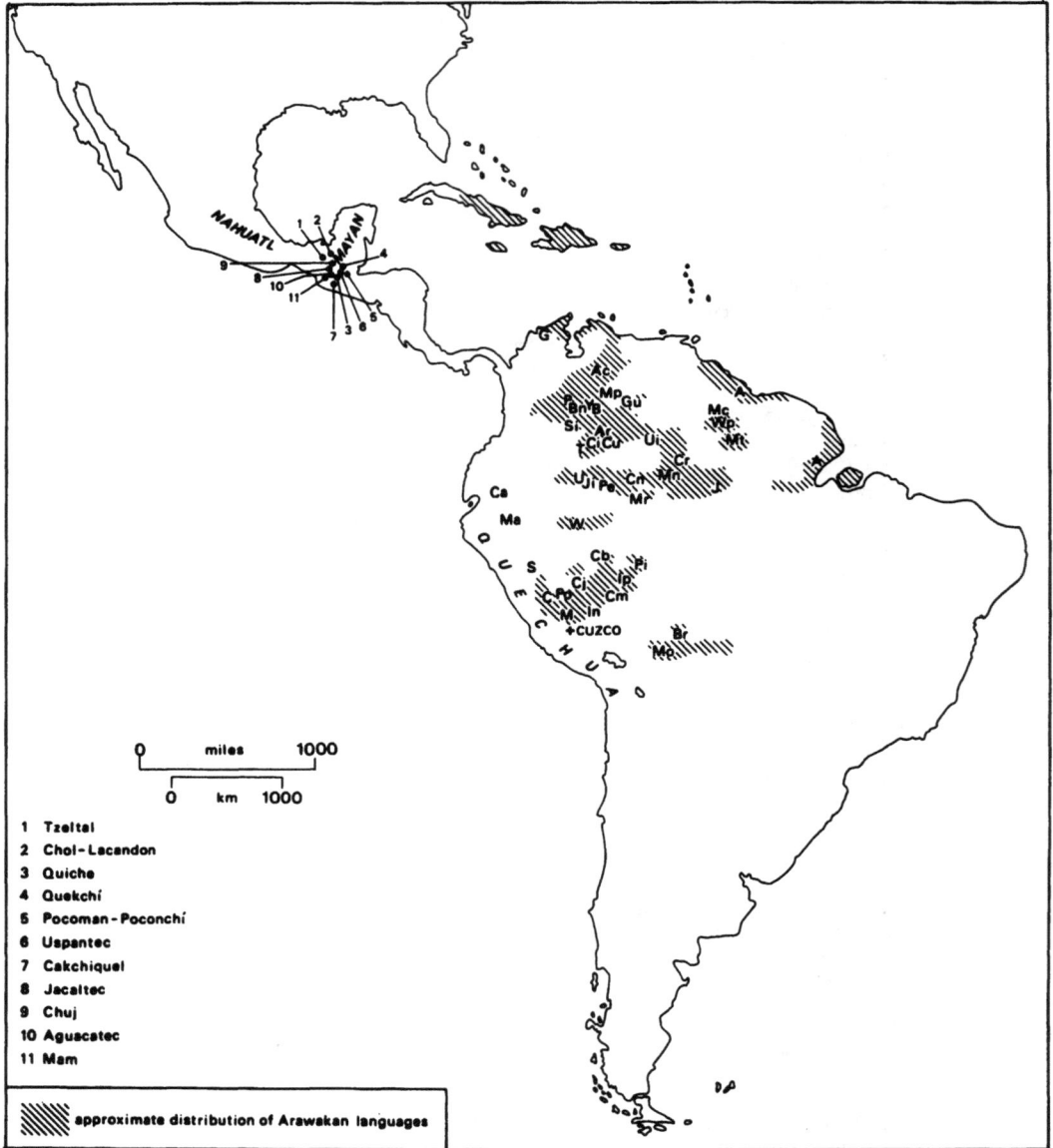

MAP 10. Tribal and linguistic distributions. A—Arawak; Ac—Achagua; Ar—Arekena; B—Baré; Bn—Baniva; Br—Bauré; C—Campa; Ca—Canelo; Cb—Cuniba; Ci—Catapolítani; Cj—Cujisenayeri; Cm—Canamari; Cn—Cayuishana; Cr—Cariaya; Cu—Carutana; G—Goajiro; Gu—Guinau; In—Inapari; Ip—Ipurina; J—Jumana; M—Machiguenga; Ma—Maina; Mc—Macusi; Mn—Manao; Mo—Mojo; Mp—Maipure; Mr—Marawa; Mt—Maopityan; P—Piapóco; Pe—Pasé; Pi—Paumari; Po—Piro; S—Shipibo; Si—Siusi; T—Tariana; U—Uainumá; Ui—Uirina; W—Waraicú; Wp—Wapishana; Y—Yavitero.

the same species may be known as *meriti*: Canamari *merity*, Inapari *meriči*, Ipuriná *merití*, *mirití*, Cuniba *möriti*, Cujisenayeri *meriti*, Paumari *myrycy*, and Piro *meriči*, *miditchi*.[122] The corresponding name for the white-lipped peccary is *irarí* (Ipuriná and Inapari), Cuniba *iyarö*, Cujisenayeri *ialö*, and Piro *iyali*, *ilavi*, *xihari*.[123]

Tucanoan

The eastern Tucanoan-speaking peoples occupy a consolidated area between the Río Caquetá and the Río Vaupés (southeastern Colombia), with the Carijona (Cariban) to the west and mainly Arawakan tribes to the north, south, and east. I. Goldman recognized eighteen "sub-tribes."[124] Their dialects are known chiefly from the work of T. Koch-Grünberg.[125] *Yehsé* or some cognate is applied to the white-lipped peccary and the same or more usually a compound word, incorporating *yehsé*, to the collared peccary (Appendix). The suffix often takes the form of *puro*, *puru*, *peraga*, or *potiro*, which may indicate Caraban influence. Among the Siona-Secoya (western Tucanoan) the white-lipped peccary is known as *sēsé*, the collared peccary as *ya:wi*.[126]

Quechua (map 10)

In highland Quechua *sintíru* describes both species of peccary (but not the domestic pig, *khuchi*).[127] It is included in the lexicographical works of Antonio Ricardo (1586) and Diego González Holguín (1608).[128] According to Bernabé Cobo (1653), "los puercos jabalíes, llamados de los indios *cintiru*, se crían solamente en las tierras calientes y al montaña, y no en mucha cantidad."[129] At the time of the Spanish conquest Quechua was spoken throughout the central Andes and in adjacent sectors of the

---

[122] Martius, 1863: p. 477; Koch-Grünberg, 1914b: p. 78; Farabee, 1922: p. 66; Goeje, 1928: p. 226.

[123] Farabee, 1922: p. 66; Goeje, 1928: p. 226. Cf. Catukina (non-Arawak) *urirý* (Martius, 1867: 2: p. 163). According to Koch-Grünberg (1914b: p. 78), *irarí* also refers to the European pig. Other Arawakan names are listed in the Appendix. Note, in particular, Baré *arúa*, Mandauáca *arúa*, *alúa*, Marawa *arua*, *arûa*, Waraicú *alûa*; Baniva and Carutana *soâra*, *tsoâra*; Mariate *kâpéna*, Uainumá *capêna*, *cabêna*.

[124] Goldman, 1948: p. 764.

[125] Koch-Grünberg, 1913: p. 960, 1914a: pp. 170, 578, 579, 822, 1915–1916: pp. 126, 430. See also Martius, 1867: 2: p. 166 (Coretú, Cueretu); Reichel-Dolmatoff, 1971: p. 200 (Desana).

[126] Vickers, 1976: p. 284.

[127] Friederici, 1947: p. 268. The same word, used as an adjective, means "montés" (Lara, 1971: p. 258).

[128] Ricardo in Aguilar Paez, 1970: pp. 112, 126; González Holguín, 1608: p. 74 (*cintiru*). Later in the standard dictionaries, Tschudi, 1853: 2: p. 478; Middendorf, 1890: p. 776; Lira, 1944: p. 917. Markham (1908: p. 210) also gives *ituchi*, possibly following Velasco (1789), 1946: 1: p. 118; not Quechua, presumably from *cuchi*, *coche*, Spanish *cochina*.

[129] Cobo, 1956: 1: p. 358.

montaña, whence it spread eastward in colonial times.[130] Variants of *sintíru* have been reported from the Machiguenga[131] and the Campa,[132] Arawakan tribes to the north of Cuzco.

Along the *montaña* and in the eastern lowlands of Ecuador and Peru the name *huangana* (*guangana, wangana*) is also employed, perhaps more particularly of the white-lipped peccary. *Huangana* is probably of Quechua origin, but the precise derivation remains obscure.[133] Mentioned by León Pinelo and Father Cobo,[134] it was in use among the Maina (Peru/Ecuador) in the seventeenth and eighteenth centuries.[135] The Quechua-speaking Canelo (Ecuador) refer to the white-lipped peccary as *huangana*,[136] and this is apparently one of several names known to the Campa.[137] It has also been reported from the valleys of the Pachitea[138] and Pastaza[139] and the province of Loreto (Peru) generally.[140]

Middle and Central America

Nahuatl (map 10)

Alonso de Molina's *Vocabulario* (1555) gives *coyametl, pitzotl* ("puerco") and *quauhtla coyametl, quauh coyametl* ("puerco montes javalin").[141] *Pítzotl* was specifically the coati (*Nasua narica narica*), but as Bernardino de Sahagún (ca. 1570) explains, "anyone who is a great eater, intemperate, is called *peçotli*".[142] Thus it might be applied to the (collared) peccary,[143] properly called *quauhcoyámetl* ("forest peccary"). Francisco Hernández

---

[130] Steward, 1948d: p. 514 (map 6).

[131] Farabee, 1922: p. 39 (*cintori*).

[132] Weiss, 1972: p. 193 (*shintóri*).

[133] Not in Ricardo (1586), 1970; González Holguín, 1608; Tschudi, 1853; Middendorf, 1890; Lira, 1944 (all apparently based on highland vocabularies); included by Santamaría (1942: 2: p. 103, collared peccary), but not Friederici, 1947. Martius (1867: p. 295) has *uankana* (as well as *sintiru*) — Quechua, "white-lipped peccary." In a study of the Shipibo (eastern Peru), R. Campos (1977: p. 59) refers to *huangana* (Castellano, white-lipped peccary), but as such it can only be a loan word.

[134] León Pinelo (1650), 1943: 2: p. 51 (provincia de Pacamoros); Cobo (1653), 1956: 1: p. 364 ("En las provincias de Tierra Firme se dice a este animal *zahino*, y en este reino del Perú lo llaman sus naturales *guangana*").

[135] Jiménez de la Espada (ed.), 1965: 3 (4): p. 246 (*guangana*, 1619); Veigl (1768), 1785: p. 199 ("*Guangana*, oder wilden Schweine"). Cf. Velasco (1789), 1946: 1: p. 118 (*guangana*).

[136] Whitten, 1976: p. 68. The Canelo adopted Quechua from the close of the sixteenth century.

[137] Denevan, 1972: p. 179.

[138] Pierret and Dourojeanni, 1966: p. 273 (white-lipped peccary).

[139] Landerman, 1973: p. 53 (Quechua, collared peccary).

[140] Paz Soldan, 1862: pp. 532, 686. See also Cabrera and Yepes, 1940: p. 280 (collared peccary); Perry, 1970: p. 39 (white-lipped peccary).

[141] A. de Molina, 1970.

[142] Sahagún, 1963: p. 10; identified by Martín del Campo, 1941: p. 496.

[143] See Dugès, 1869: p. 138 (*quapicotl*); Starr, 1902: p. 78 (*pizotole*); Brewer and Brewer, 1962: p. 55 (*cuajpitzotl*).

(1571–1576) has "el coyámetl, que algunos llaman *quauhcoyámetl* porque es montés, otros *quauhtla coyámetl* y otros *quauhpezotli.*"[144]

Macromayan (map 10)

Citam (*quitam*, *chitam*, hispanized to *guitame*) is applied, usually with some adjectival qualification, to both peccaries and to the European pig.[145] The earliest known reference (*quitan*) is in a *relación* of Sucopo (near Valladolid, northern Yucatán) of 1579.[146] In Tzeltal (of Tenejapa, highland Chiapas), according to E. S. Hunn, *čitam* is "pig," *wamal čitam* "bush pig" or peccary, with up to four subdivisions of the latter, probably "corresponding to the adults and juveniles of the two species."[147] In Yucatán and adjacent parts of Chiapas *čitam* may refer particularly to the collared peccary,[148] *keken* (*kekem*, *qeqem*) to the white-lipped peccary.[149] A late seventeenth-century account of Chol-Lacandon (Chiapas) gives *cehcem.*[150] In Quiché, Quekchí, Pocoman-Poconchí, Uspantec (central highlands of Guatemala), and Cakchiquel (to the south), the pig is known as *ak* (*'aq*, *'ahq*, *aj'k*), the peccary as *quiché ak* ("bush pig").[151] In Aguatec and Mam *ak* becomes *boch.*[152]

Chibchan and Misumalpan (map 8)

The languages of Central America, south and east of the lands of Mayan occupation, include two important families, Chibchan (Panama, Costa Rica) and Misumalpan (Nicaragua, part of Honduras). In these languages peccaries are known by a considerable number of names (Appendix), with similarities within and between the families. In Tala-

---

[144] Hernández, 1959: 1: pp. 310–311; identified by Lichtenstein, 1830: p. 114. Cf. Seler, 1909: p. 402; Clark (ed. *Codex Mendoza*, ca. 1580), 1938: 1: p. 8 (*coyametl*), 2: pp. 54, 77; Carrión (1581), 1965: p. 99 (*coyame[tl]*). For names applied in some other languages of Mexico, see Appendix.

[145] Stoll, 1884: p. 56.

[146] Latorre (ed.), 1920: p. 85.

[147] Hunn, 1977: pp. 225–227.

[148] Stoll, 1887: p. 141 (Jacaltec, *belajil chitam*; Chuj, *c'ultquil chitam*); Noyes (ed.) in Ponce, 1932: p. 308; Hatt, 1938: p. 336; Redfield and Villa Rojas (1934), 1962: p. 379; Slocum and Gerdel, 1965: p. 65 (Tzeltal of Bachajon, *ha'mal chitam*).

[149] Seler, 1909: p. 402; Murie, 1935: p. 28; Slocum and Gerdel, 1965: p. 65 (*q'uem*). Gaumer (1917: p. 67) has *cehuikax*.

[150] Hellmuth, 1977: p. 430, quoting Francisco Moran *Arte en lengua Cholti* (ms. 1695). Becerra (*Chol*, 1937: pp. 264, 271) gives (a) jabalí (white-lipped peccary), *muk-chitiam*, *matié-chitiam*, (b) puerco montés, *maltik-chitiam*, *matiel-chitam*.

[151] Brinton, 1884: p. 95; Stoll, 1884: p. 56, 1888: p. 184; Seler, 1909: p. 402; Fernández, 1937–1938: p. 187 (*quiché-ajk*); Ibarra, 1959: p. 169; Kaufman, 1964: p. 94 (*'ahq*); Edmonson, 1965 [Dict.] (*'aq*).

[152] Stoll, 1884: p. 56, *stzelá boch, boch*, 1887: p. 141, *stsela boch, xboch*; Reynosa (1644), 1897: p. 331 (*tzelah ixbocx*). In outlying areas of Macromayan, Gulf Coast Huasteca and Totonac, Mixe of southern Oaxaca, and Popoluca of southern Vera Cruz, different names are applied (Appendix).

manca the white-lipped peccary is known as *sené*,[153] in Bri-Bri *siní*,[154] in Guaymí (Changuena, Chumula, Gualaca) *siri*,[155] in Cabécar *sir-bí*,[156] in Tirribí and Terraba *shirí*[157] (all Chibchan); in Sumo *sévi*,[158] in Ulva *siuí*[159] (Misumalpan); and in Subtiaba (Hokan, northwest Nicaragua) *siñúi*.[160] Other subdivisions of Guaymí have *motú-kri* (Move), *metu-kri* (Norteño), *motoro-kri* (Penonomeño).[161] The collared peccary is *ka'-sir* or *kas'-ri* among the neighboring Bri-Bri and Cabécar.[162] The Miskito (Misumalpan) call the species *búksa*[163] (whence the Anglicized *bookser*), the Rama (Chibchan) *múksa*,[164] to which are perhaps related Sumo *mulkus*, *múlucus*, and Ulva *môlōkōs*.[165]

---

[153] Lehmann, 1920: 1: p. 333.

[154] Ibid., p. 319; Gabb, 1876: p. 564 (*si-ni*). Cf. Boruca *si-ni'*, collared-peccary, according to Gabb, 1876: p. 564.

[155] Lehmann, 1920: 1: p. 164.

[156] Gabb, 1876: p. 594; 1883: p. 480.

[157] Gabb, 1876: p. 594 (*shi-ri'*); Lehmann, 1920: 1: p. 266.

[158] Ibid., pp. 508 (*sivĕ*), 578. *Sauakaya* in Conzemius, 1932: p. 166.

[159] Lehmann, 1920: 1: pp. 508, 571 (*saui, sowie*).

[160] Ibid., p. 944.

[161] Ibid., p. 164.

[162] Gabb, 1876: pp. 564, 594; Lehmann, 1920: 1: p. 319.

[163] Ibid., p. 508; Conzemius, 1927: p. 347; T. Young, 1842: p. 170.

[164] Lehmann, 1920: 1: p. 456; Conzemius, 1927: p. 347.

[165] Lehmann, 1920: 1: pp. 508, 574, 576, 578; Conzemius, 1927: p. 347.

D: HUNTING

Methods and equipment (map 11)

Traps

Peccaries may be captured alive, and then either killed or the juveniles taken to be tamed, reared, and fattened. Among tribes living between the estuary of the Orinoco and northern Brazil (Waiwai, "old Warrau," a few Caribs of the Pomeroon coast,[1] Macusi,[2] and Wapishana[3]), conical cages resembling fishing creels are placed over the end of a hollow log or the entrance to a hole in the ground (such as the burrow of the giant armadillo) where a peccary—almost invariably the collared peccary—has taken refuge (fig. 7). The animal is then allowed to leave the hole or is forced out by using pointed stakes or torches.[4] The Makú (Colombia) build a corral around the hole, placing dead wood and tying vines between adjacent trees.[5]

The ancient Maya trapped peccaries in a kind of noose, as shown in the *Codex Tro-Cortesianus* (p. 34 supra). It is likely that the method was more widely employed. The Cáhita (Mayo, of northwestern Mexico) set rope snares for deer and peccary that are "possibly aboriginal."[6]

Drives and the use of nets were reported by Pascual de Andagoya (1514–   ) from Coiba, western Panama (territory of the Guaymí). Here peccaries "were caught with large nets of stuff like hemp, called by the Indians *nequen* [henequen], the meshes being a finger in breadth. These nets were fastened at the entrance of a wood where there was a herd of peccary, which came against the nets and were unable to get through the meshes. Then the people called out, the nets fell over the peccaries, and they were killed with lances. . . ."[7] The method, accompanied at times by the firing of tall grass, implies a more open, parkland landscape than exists at present.[8] According to Pietro Martire d'Anghiera (1516),

---

[1] Roth, 1924: p. 183 (the lower Pomeroon was mainly occupied by Arawak).

[2] Farabee, 1924: p. 43.

[3] Farabee, 1918: p. 52.

[4] Peccaries hunted by the Cáhita (north-western Mexico) "were sometimes smoked out of dens in the rocks" (Beals, 1943: p. 13). H. Sick (1959: p. 43) and K. I. Taylor (1972: p. 14) also refer to the use of smoke. A Mundurucú (Brazil) myth relates how a band of collared peccary (*catitú*) entered a hole, which was then blocked, and one animal released and killed daily (Murphy, 1958: pp. 86–87).

[5] Silverman-Cope, 1973: p. 73 (collared peccary).

[6] Beals, 1945a: p. 12.

[7] Andagoya, 1865: p. 18.

[8] Lothrop, 1950: p. 13; Sauer, 1966: pp. 244, 274, 286; Torres de Araúz, 1977: pp. 69–70.

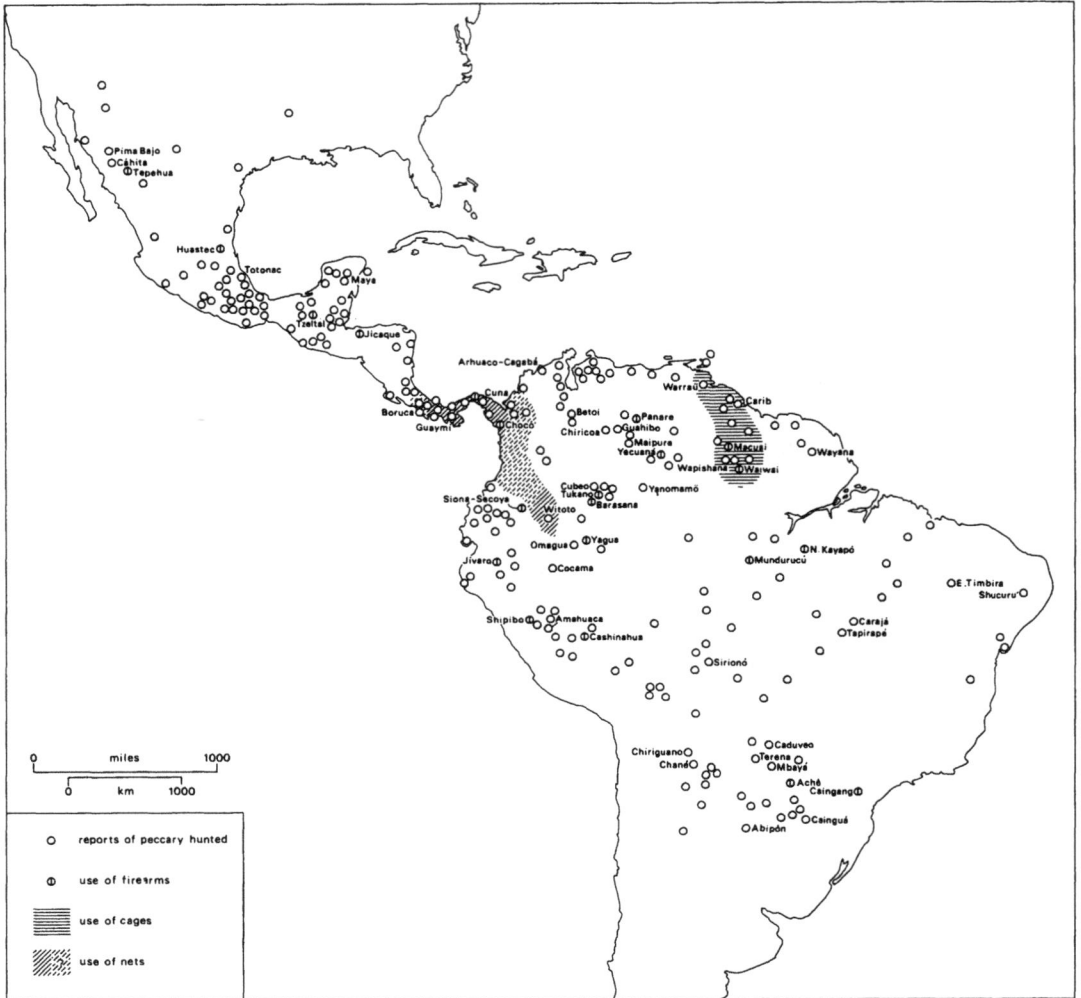

MAP 11. Hunting peccary: use of cages, nets, and firearms.

jabalíes were also caught in pits (*fosos*), dug along trails and disguised with branches (*ramaje*).[9] Fernández de Oviedo y Valdés (1514–1526) echoed Andagoya in observing that the Indians of *Tierra Firme* (probably the province of Cueva, Cuna territory, to the east of Coiba) "trap deer and pigs with branches and . . . nets, into which the animals fall. At times they hunt and beat them out, and with a great number of people they attack them and take those that they kill with arrows [*saetas*, more correctly 'darts'] and spears."[10]

---

[9] Anghiera, 1944: p. 298. Anghiera's observation was based on expeditionary reports by Francisco Becerra and Gonzalo de Badajoz in 1514.

[10] Fernández de Oviedo y Valdés (1526), 1950: pp. 117–118, 152 (*cepos*, traps for

FIG. 7. A peccary trap, Waiwai, Guiana. W. Roth, 1924: pl. 44.

Drives towards carefully positioned nets may have been the preferred method in lower Central America.[11] The only known report for South America concerns the Witoto (eastern Peru), who "capture peccaries, deer and tapirs in a great net, six feet high and a thousand or fifteen hundred feet long, which is stretched among the trees in a suitable place in the forest."[12]

At the end of the nineteenth century the Caingúa (northern Argentina) used peccary "traps."[13] In Paraguay, according to Félix de Azara (ca. 1800), "the natives observe the paths [taken by collared peccary], and forming a long hedge of branches on each side, with a deep pit at the end, they frighten them by hallooing . . . and drive them towards the ditch, which is generally filled with them."[14] One of the hunting rituals performed by the Mundurucú (Brazil) "simulates the use of a runway of stakes to trap peccaries."[15] The Shucuru' (eastern Brazil),[16] the Terena and the Mbayá (Brazil-Paraguay)[17] also prepare concealed pitfalls, and this simple procedure is probably widely employed in the Selval region.[18]

Men and dogs

Peccaries are hunted both by individuals and by groups of men, occasionally assisted by dogs.[19] Some tribes do not appear to have possessed (hunting) dogs when first contacted;[20] elsewhere they have probably been acquired comparatively recently.[21] When dogs are said to

---

peccaries), 1959a: pp. 29, 50–51. Trimborn (1952: p. 257) refers to "flint-bladed missiles hurled from slings"; see also Sauer, 1966: p. 244. Cf. Steward, 1948a: p. 31 (Cuna). According to D. B. Stout (1947: p. 22), the present San Blas Cuna use shotguns, not traps.

[11] Armas, 1888: p. 69.

[12] Farabee, 1922: p. 138.

[13] Ambrosetti, 1894a: p. 729.

[14] Azara, 1838: p. 119.

[15] Horton, 1948: p. 273. Farabee (1917: pp. 135–136) also refers to a "pig trap" (upper Amazon).

[16] Hohenthal, 1954: p. 112. Hans Staden ([1547–1555], 1874: p. 160) remarked that the peccary was "very difficult to catch in traps, which the savages [of eastern Brazil] use for the purpose of catching game."

[17] Métraux, 1946a: p. 257; Oberg, 1949: p. 10. See also Armas, 1888: p. 70.

[18] Cooper, 1949: p. 273.

[19] Brettes, 1903: p. 340 (Arhuaco-Cagabá, Colombia); Nordenskiöld, 1912: p. 48 (the Chaco), 1920: p. 31 (Chiriguano, southern Bolivia, Chané, northern Argentina); Farabee, 1918: p. 51 (Wapishana, Guiana-Brazil—both species of peccary), 1924: pp. 43, 51–54, 155 (Macusi, Waiwai, Wapishana, Guiana-Brazil); Schomburgk (1840–1844), 1923: 2: p. 129 (Guiana, ? Macusi); Roth, 1924: pp. 182–183 (Guiana); Wafer (1680–1688), 1934: p. 21 (Darién); Karsten, 1935: p. 173 (Jívaro, Ecuador); Stirling, 1938: p. 105 (Jívaro); Horton, 1948: p. 280 (Mundurucú, Brazil); Lothrop, 1948: p. 253 (Panama [Coiba]); Métraux, 1948b: p. 451 (Caingang, Brazil); Dreyfus, 1963: p. 29 (Northern Kayapó, Brazil); Henry, 1964: pp. 100, 157 (Caingang); Hurault, 1968: p. 6 (Wayana, French Guiana); Murphy and Murphy, 1974: p. 63 (Mundurucú).

[20] Nimuendajú, 1939: p. 94 (Apinayé); Lowie, 1946b: p. 482 (Ge); McKim, 1947: p. 99 (Cuna); Carneiro, 1974: p. 132 (Amahuaca). The Sirionó had no dogs in the late 1940s (Holmberg, 1950: p. 29).

[21] Nimuendajú, 1946: p. 75 (Eastern Timbira); K. I. Taylor, 1972: p. 15 (Sanumá [Yąnomamö]). Seldom employed by the Boruca of Costa Rica (Stone, 1949: p. 8) or by the Cashinahua of eastern Peru (Kensinger, 1975: p. 28, except in hunting some small animals).

be used and the species of peccary is named (a minority of observations), the less gregarious but also more predictable collared predominates.[22] Dogs cut out and surround stragglers. Droves of white-lipped peccaries, on the other hand, make a considerable noise and may also be detected some distance away by their characteristic odor. Dogs would jeopardize a silent approach.[23] The Makú (Colombia) bind their muzzles with vines to prevent barking.[24] The "only value [of dogs] in attacking white-lippeds may be as 'bait,' that is, when they come rushing back to their masters with peccaries charging at their heels."[25] Much will therefore depend on whether dogs are regarded as of much assistance in hunting generally and on whether the hunt has been organized with peccaries, of one or other or both species, in mind.[26]

The Djuka, bush Negroes of Surinam, "dress" or prepare their dogs by forcibly administering (sometimes through the nostrils) an infusion of the dried meat of the animal to be hunted, especially the peccary.[27] The Jívaro (Ecuador) anoint a new hunting dog with the blood of the first peccary that is killed.[28] Among the Waiwai and the Wapishana the dogs' senses are sharpened by having their muzzles "smeared with pepper or with evil-smelling *apoporé* bark," or they may be given a decoction of *apoporé* bark to drink.[29] The former is also practiced by some Amazonian settlers who rub the noses of their dogs with the leaves of *Piper lanceolatum*.[30] The Siona-Secoya (eastern Ecuador) feed their dogs a mixture of plantains boiled with the magical leaves of *Xanthosoma* sp. (the latter supplied by shamans) prior to hunting the collared peccary.[31]

The Shipibo,[32] the Machiguenga,[33] and the Cashinahua[34] of eastern Peru normally hunt alone, the Cashinahua combining only to attack tapir and peccary. The Cubeo (southeast Colombia) "until recently"

---

[22] Alston, 1879–1882: p. 108 (Vera Paz, Guatemala); E. A. Goldman, 1920: p. 74 (Panama); Miller, 1930: p. 18 (southern Mato Grosso); Beals, 1943: p. 13 (Cáhita, northwest Mexico, outside the range of white-lipped peccary); Kelly and Palerm, 1952: p. 74 (Tajin Totonac, outside the present range of white-lipped peccary); Gilmore (1950), 1963: p. 382; Yde, 1965: p. 122 (Waiwai, Guiana-Brazil, usually not used in hunting white-lipped peccary); Silverman-Cope, 1973: p. 73 (Makú, Colombia); Vickers, 1976: p. 118 (Siona-Secoya, Ecuador); Pennington, 1979–1980: 1: p. 211 (outside the range of white-lipped peccary); Coe and Diehl, 1980: 2: p. 102 (Río Chiquito, Vera Cruz, Mexico).

[23] Wilbert, 1972: p. 42 (Yąnomamö [Yanoama], Brazil-Venezuela); Smith, 1976: p. 456 (Amazonian settlers). Cf. Chagnon, 1968: p. 29.

[24] Silverman-Cope, 1973: p. 74.

[25] Kiltie, 1980: p. 543 n. 12.

[26] K. I. Taylor (1972: p. 16) refers to a dog that "specialized in collared peccary, paca and agouti" (Sanumá [Yąnomamö]).

[27] Kahn, 1931: pp. 81–82.

[28] Karsten, 1935: p. 173.

[29] Yde, 1965: pp. 120, 122.

[30] Smith, 1976: p. 456.

[31] Vickers, 1976: p. 118.

[32] Campos, 1977: p. 60.

[33] A. Johnson, 1977: p. 159.

[34] Kensinger, 1975: p. 27.

hunted peccaries in groups.[35] Parties of the Siona-Secoya[36] and the northern Aché (eastern Paraguay) pursue the white-lipped peccary, but not the smaller species.[37] White-lippeds are "one of the few [animals] that are sometimes hunted co-operatively" by the Sirionó (eastern Bolivia).[38] Similarly, the Northern Kayapó (Brazil)[39] and the Amahuaca (eastern Peru)[40] band together to hunt the larger species, but no other animal. Members of some tribes hunt "peccaries" both alone and cooperatively,[41] but the latter is more frequently reported.[42] A combined expedition may be organized after a herd has been sighted by a lone hunter. "Peccaries evidently forage slowly enough that they will still be in the neighborhood the next day, so that a proper hunt can be well planned."[43] Parties may consist of only two or three (often related) individuals[44] or much larger groups, sometimes all the adult males of entire communities.[45] When white-lipped peccaries are reported, a communal drive is almost invariably organized. Tracking may occupy several days. Wherever possible, herds are approached against the wind,[46] then surrounded. Peccaries are most vulnerable when driven into water,[47] or when caught crossing rivers (fig. 8),[48] and when concentrated on islands

---

[35] I. Goldman, 1963: p. 10.

[36] Vickers, 1976: p. 97.

[37] Hawkes, Hill, O'Connell, 1982: p. 383 (the Capuchin monkey and the paca are also hunted cooperatively).

[38] Holmberg, 1950: p. 25.

[39] J. B. Turner, 1967: p. 104.

[40] Carneiro, 1974: p. 124.

[41] Stout, 1948a: p. 257 (Cuna, Panama); Oberg, 1949: p. 10 (Caduveo, Brazil); Redfield and Villa Rojas, 1962: p. 48 (Maya, Yucatán, Mexico); Henry (1941), 1964: p. 100 (Caingang, Brazil); Chagnon, 1968: p. 29 (Yanomamö, Brazil-Venezuela).

[42] Dobrizhoffer (1784), 1822: 1: p. 270 (Abipón, Paraguay); Farabee, 1924: p. 43 (Macusi, Guiana-Brazil); Gillin, 1936: p. 3 (Carib, R. Barama, Guiana); Stirling, 1938: p. 105 (Jívaro, Ecuador); Fejos, 1943: pp. 40–41 (Yagua, eastern Peru); Beals, 1945a: p. 12 (Cáhita [Mayo], ? Yaqui, northwest Mexico); Métraux, 1946b: p. 451 (Caingang, Brazil); Stout, 1947: p. 22 (San Blas Cuna, Panama); Kirchhoff, 1948c: p. 448 (Guahibo and Chiricoa, Venezuela); Lipkind, 1948: p. 181 (Carajá, Brazil); Steward and Métraux, 1948b: p. 730 (Yagua, eastern Peru); Murphy, 1960: p. 54 (Mundurucú, Brazil); Barandiaran, 1962: pp. 15–16 (Yecuaná, Venezuela); Dreyfus, 1963: p. 29 (Northern Kayapó, Brazil); Frikel, 1968: p. 94 (Xikrín: Northern Kayapó); Wilbert, 1972: p. 42 (Yanomamö [Yanoama], Brazil-Venezuela); Kaplan, 1975: p. 38 (Piaroa, Venezuela); Dumont, 1976: p. 62 (Panare, Venezuela); Wagley, 1977: p. 62 (Tapirapé, Brazil).

[43] Beckerman, 1980: pp. 94–95 (Barí, Venezuela).

[44] Brettes, 1903: p. 340 (Arhuaco-Cagabá, Colombia; rarely bands); Farabee, 1918: p. 54 (Wapishana, Guiana-Brazil); Stone, 1949: p. 8 (Boruca, Costa Rica).

[45] Stirling, 1938: p. 105 (Jívaro, Ecuador); Murphy and Murphy, 1974: p. 63 (Mundurucú, Brazil).

[46] Holmberg, 1950: p. 25.

[47] Stirling, 1938: p. 105 (Jívaro, Ecuador); Métraux, 1946a: 257 (Mbayá, Paraguay-Brazil).

[48] Crévaux, 1883: pp. 51–54; Thurn, 1883: pp. 54–55, 109; Kappler, 1887: p. 80; Schomburgk (1840–1844), 1923: 2: p. 130; Sick, 1959: p. 155 (Rio Tapajós, Brazil). Perry (1970: p. 45) refers to reports of one to two thousand *huanganas* (white-lipped peccaries) swimming across rivers.

Fig. 8. A band of white-lipped peccaries, attacked as they swim across the River Marowijne, Surinam. A. Kappler, 1887: p. 80.

during the wet season.[49] In such circumstances substantial numbers are likely to be slaughtered.

Weapons

The lance and the bow and arrow are the principal traditional weapons used in hunting peccary. Blowguns have only rarely been reported (Jívaro,[50] Yagua[51]). According to José Gumilla (ca. 1745), Indians living along the (? lower and middle) Orinoco employed, besides the bow and arrow, barbed "harpoons" (arpones) of bone or iron, securely attached to a shaft by a cord. On impact, the shaft was thrown free, but quickly got caught in the undergrowth.[52] This was the "lance tied to cords" of the Maipure, mentioned by Humboldt (1799–1800).[53] The "Waiwai (Guiana-Brazil) use a detachable bamboo-headed arrow for the same purpose and with similar results."[54] According to Jens Yde (1965), the Waiwai hunt peccary "with the bamboo-bladed arrow orahnó, with the harpoon-arrow tarurüká, and with the harpoon-like arrow yóchopotóro."[55] Harpoon-like metal points are also employed by the Barí (Colombia).[56] Before the introduction of firearms, the Siona-Secoya (eastern Ecuador) used lances with detachable points against peccary and tapir.[57]

The bow and arrow[58] is reported rather more frequently than the lance.[59] Some tribes use both.[60] A preference for one or the other may

---

[49] Lipkind, 1948: p. 181 (Carajá, Brazil); Wagley, 1977: p. 62 (Tapirapé, Brazil).

[50] Stirling, 1938: p. 105 (together with other weapons).

[51] Steward and Métraux, 1948b: p. 730. The blowgun belongs chiefly to the northern and western basin of the Amazon/Orinoco (Yde, 1948: p. 305 [map]).

[52] Gumilla, 1791: 1: p. 257.

[53] Humboldt (1799–1804), 1852–1853: 2: p. 269.

[54] Roth, 1924: pp. 182–183.

[55] Yde, 1965: p. 122.

[56] Beckerman, 1980: p. 92 (against tapir, peccary, and bear).

[57] Vickers, 1976: p. 97.

[58] Lozano, 1733: p. 40, 1873–1874: 1: pp. 286–288 (La Plata, Chaco); Thurn, 1883: p. 241 (Guiana); Koch-Grünberg, 1909–1910: 1: p. 103 (Río Aiarí); Krause, 1911: p. 386 (Kayapó); Schomburgk (1840–1844), 1923: 2: p. 129 (Guiana, ? Macusi); Roth, 1924; pp. 182–183 (Waiwai, Guiana-Brazil); Wagley and Galvão, 1948b: p. 169; Hernández de Alba, 1948c: p. 394 (Betoi, Colombia); Holmberg, 1950: pp. 14, 25, 32 (Sirionó, Bolivia); Barandiaran, 1962: p. 15 (Yecuaná, Venezuela); Dreyfus, 1963: p. 29 (Northern Kayapó, Brazil); Yde, 1965: p. 122 (Waiwai); Pierret and Dourojeanni, 1967: p. 18 (lower Río Ucayalí); Frikel, 1968: p. 93 (Xikrín [Northern Kayapó], Brazil); Chagnon, 1968: pp. 29, 32 (Yanomamö, Brazil-Venezuela); Ruddle, 1970: p. 41 (Maracá, Colombia-Venezuela); Wilbert, 1972: p. 42 (Yanomamö); Cordova-Rios and Lamb, 1972: pp. 54–55 (Amazon region); Silverman-Cope, 1973: p. 73 (Makú, Colombia; white-lipped peccary); Carneiro, 1974: pp. 123–124 (Amahuaca, Peru); Murphy and Murphy, 1974: p. 63 (Mundurucú, Brazil); Kensinger, 1975: p. 27 (Cashinahua, Peru); Wagley, 1977: p. 67 (Tapirapé, Brazil); Campos, 1977: pp. 60–61 (Shipibo, Peru); Hames and Vickers, 1982: p. 371 (Yanomamö); Hawkes, Hill and O'Connell, 1982: p. 383 (Aché, eastern Paraguay).

[59] Dobrizhoffer (1784), 1822: 1: p. 270 (Abipón, Paraguay); Humboldt (1799–1804), 1852–1853: 2: p. 269 (Maipure, Guiana-Venezuela); Andagoya (1514–1546), 1865: p. 24 (in Coiba, Panama); Stirling, 1938: p. 105 (Jívaro, Ecuador): Métraux, 1948c: p. 692 (Cocama and Omagua, Peru-Brazil); Fernández de Oviedo y Valdés (1514–1526), 1950: p. 152, 1959a: p. 51; Le Roy Gordon, 1957: p. 23 (Chocó, Colombia); I. Goldman, 1963: p. 10 (Cubeo, Colombia); Villa Rojas, 1969b: p. 206 (Tzeltal, Chiapas, Mexico); Vickers, 1976: p. 97 (Siona-Secoya, Ecuador); Henley, 1982: p. 44 (Panare, Venezuela).

[60] Gumilla (ca. 1745), 1791: 1: pp. 257–260 (along the Orinoco); Farabee, 1918: pp. 51–54 (Wapishana, Guiana-Brazil), 1924: p. 43 (Macusi, Guiana-Brazil); Wafer (1680–1688),

sometimes be inferred, but the available information is usually neither explicit nor unequivocal. Peccaries caught in nets (lower Central America and locally elsewhere) were presumably dispatched with lances or clubs. The use of poisoned arrows appears to be exceptional (Makú,[61] Yąno-mamö,[62] Yagua,[63] Macusi[64]). In hunting the white-lipped peccary, the Sirionó "take with them only their bamboo-headed arrows (*tákwa*), as only these are effective in killing such a large animal."[65] Traditional weapons, of whatever kind, have the advantage of local manufacture. They are employed with great skill, based on long experience, and are also virtually silent—and correspondingly effective against animals found in herds. Again, to hunt with lance or arrow may be regarded as "more thrilling"[66] and possibly more manly. Nevertheless, firearms of various kinds have been adopted since at least the middle of the eighteenth century[67] and are now widely reported in use against the peccary.[68] They appear to be employed more by lone hunters than by parties of hunters, and in areas remote from European influence they have rarely (if at all) entirely displaced traditional weapons.

Products of the chase

Meat (map 12)

For many aboriginal groups—whether primarily hunter-gatherers or horticulturists—peccaries are (or have been) among the principal game animals. They have probably contributed more to human diet than any

---

1934: p. 64 (Darién); Fejos, 1943: pp. 40–41 (Yagua, Peru); Beals, 1945a: p. 12 (Cáhita [Mayo]); Métraux, 1948b: p. 451, Henry, 1964: pp. 100, 157 (Caingang, Brazil); Pennington, 1969: p. 128 (Tepehua, Mexico).

[61] Silverman-Cope, 1973: p. 73 (white-lipped peccary, paralyzed within about 100 meters).

[62] Wilbert, 1972: p. 42. Cf. Koch-Grünberg, 1909–1910: 1: p. 103.

[63] Fejos, 1943: p. 41.

[64] Farabee, 1924: p. 43.

[65] Holmberg, 1950: p. 25.

[66] Henley, 1982: p. 44 (Panare, Venezuela).

[67] Lozano, 1873–1874: 1: p. 288 (*escopeta*, in the region of La Plata and the Chaco).

[68] Crévaux, 1883: pp. 51–54 (French Guiana); Schomburgk (1840–1844), 1923: 2: p. 129 (Guiana, ? Macusi); Stirling, 1938: p. 105 (Jívaro, Ecuador); Von Hagen, 1943: p. 49 (Jicaque, Honduras); Fejos, 1943: pp. 40–41 (Yagua, Peru); Stout, 1947: p. 22 (Cuna, Panama); Fulop, 1954: p. 101 (Tukano, Colombia); Barandiaran, 1962: p. 15 (Yecuaná, Venezuela); Dreyfus, 1963: p. 29 (Northern Kayapó, Brazil); Henry, 1964: pp. 100, 157 (Caingang, Brazil); Yde, 1965: p. 122 (Waiwai, Guiana-Brazil); J. B. Turner, 1967: p. 105 (Northern Kayapó); Pierret and Dourojeanni, 1967: p. 18 (lower Río Ucayali); Frikel, 1968: p. 93 (Xikrín [Northern Kayapó], Brazil); Bennett, 1968b: p. 41 (Chocó, Panama); Pennington, 1969: p. 128 (Tepehua, Mexico); Laughlin, 1969: p. 300 (Huastec, Mexico); Murphy and Murphy, 1974: p. 63 (Mundurucú, Brazil); Langdon, 1975: p. 15 (Barasana and Taiwano, Colombia); Kensinger, 1975: p. 28 (Cashinahua, Peru: 1955, only traditional weapons; 1959, four men with shotguns; 1963, all men with shotguns); Campos, 1977: pp. 60–61 (Shipibo, Peru); Ross, 1978: p. 7 (Achuará Jívaro); S. Hugh-Jones, 1979: p. 30 (Barasana, Colombia); Vickers, 1976: p. 98, 1979: pp. 20, 22 (Siona-Secoya, Ecuador); Henley, 1982: p. 44 (Panare, Venezuela); Hames and Vickers, 1982: p. 371 (Yecuaná); Hawkes, Hill and O'Connell, 1982: p. 383 (Aché, eastern Paraguay).

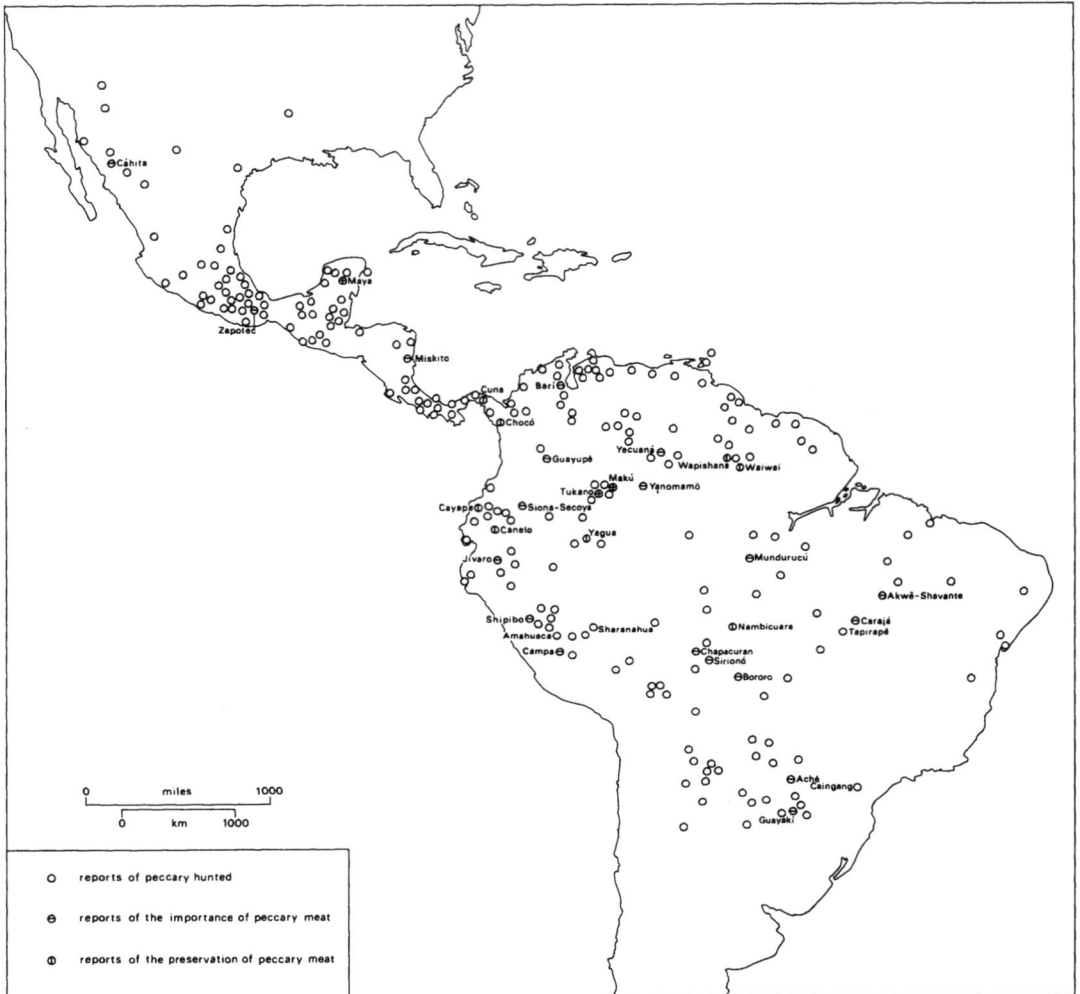

MAP 12. Peccary meat.

other terrestrial species. Lionel Wafer (1680–1688) observed that the
pecary (collared) and the *warree* (white-lipped) were the "chief game" of
the Indians of Darién.[69] Similarly, peccaries are known to have been
"important" to, or a "favourite food" of, the Campa (Peru),[70] the Sirionó
(Bolivia),[71] the Bororo (Brazil-Bolivia),[72] the Chapacura (Brazil-Bolivia),[73]

---

[69] Wafer (1699), 1934: p. 102.
[70] Denevan, 1972: p. 171; Weiss, 1972: p. 193.
[71] Holmberg, 1950: pp. 24–25.
[72] Lowie, 1946a: p. 420.
[73] Métraux, 1942: p. 88, 1948a: p. 399.

the Akwẽ-Shavante (Brazil),[74] the Carajá (Brazil),[75] the Caingang (Brazil),[76] the Tukano (Colombia),[77] the Guayupé (Venezuela-Colombia),[78] the Maya (Mexico-Guatemala),[79] the Cáhita,[80] and the Zapotec (Mexico).[81] In Guiana, according to R. H. Schomburgk, they were "more productive than . . . any other animal of the chase."[82] Over a period of one month in a community of the Achuarä Jívaro (Ecuador), the two species accounted for 60 percent by weight of the total harvest of animals.[83]

The white-lipped peccary has been described as "the principal game food" (Mundurucú, Brazil),[84] "the most useful animal" (Guayakí, Paraguay),[85] and "the main stock of meat" (in Tortuguero, Caribbean coast of Costa Rica).[86] According to R. B. Hames, it is "the most important game [animal] for the majority of neotropical hunters."[87] A number of calculations of the percentages of meat contributed by game animals to the diet of various aboriginal groups put the white-lipped peccary in first or second place and above the collared peccary. Such calculations refer to communities of the Barí (northern Colombia-Venezuela),[88] the Yạnomamö (Venezuela-Brazil),[89] the Yecuaná (Venezuela),[90] the Miskito (Nicaragua),[91] the Makú (Colombia),[92] the Siona-Secoya (Ecuador),[93] the

[74] Maybury-Lewis, 1967: p. 37.
[75] Lipkind, 1948: p. 181.
[76] Henry (1941), 1964: p. 157.
[77] Fulop, 1954: p. 101 (among "los animales mas codiciados"). Woolly monkeys and peccaries are "the most esteemed game" of the Barasana of Colombia (S. Hugh-Jones, 1979; p. 30).
[78] Kirchhoff, 1948a: p. 386.
[79] Olsen, 1982: p. 9.
[80] Beals, 1943: p. 13.
[81] Nader, 1969a: p. 339.
[82] Schomburgk (1840–1844), 1923: 2: p. 129.
[83] Ross in Lizot, 1979: p. 152.
[84] Murphy, 1960: p. 54.
[85] Cadogan, 1973: p. 102.
[86] Frost, 1974: p. 159.
[87] Hames, 1980: p. 42.
[88] Beckerman in Ross, 1978: p. 18 (white-lipped peccary in first place, collared peccary in third); ibid., 1980: pp. 94–95 (agouti in second place; peccaries together marginally more important [kilograms of meat per person per year] than exploited species of monkey).
[89] Lizot, 1979: p. 151 (white-lipped peccary in first place, collared peccary in third); Ross in Lizot, 1979: p. 154; Hames, 1979: pp. 232, 238. Cf. Hames and Vickers, 1982: p. 362.
[90] Hames, 1979: pp. 232, 238 (white-lipped peccary in second place). Cf. Hames and Vickers, 1982: p. 362.
[91] Nietschmann, 1972a: p. 50, 1973: pp. 83, 107, 165 (white-lipped peccary next to the turtle; the meat of the collared peccary is rejected).
[92] Silverman-Cope, 1973: p. 89 (white-lipped peccary in first place, collared peccary in second).
[93] Vickers, 1976: p. 140 [1973–1974], 1979: p. 22, 1980: p. 12 (white-lipped peccary [30.2 percent] in first place, collared peccary [19.3 percent] in second). Cf. Hames and Vickers, 1982: p. 362.

Shipibo (Peru),[94] and the Aché (eastern Paraguay).[95] In a Surinam village, over a one-month period (ca. 1970), the number of white-lipped peccaries killed was more than double that of collared peccaries.[96] A survey (1966) of 430 families along the lower Ucayalí (Peru) revealed that game contributed 23.65 percent of animal protein (fish, 61.68 percent), and of this the white-lipped peccary accounted for the largest proportion, followed by the collared peccary.[97] On the Río Pachitea (ca. 1965) more meat than fish was consumed; however, the white-lipped peccary ranked only ninth by weight (3.10 percent) and the collared peccary third (16.5 percent).[98] Likewise, the collared peccary was found to occupy a higher rank among the Sharanahua (Peru)[99] and the Amahuaca (Peru).[100]

Peccaries are generally important for several reasons. They are among the larger mammals of Neotropica, and the combined distribution of the two species is very extensive. Their overall density is comparatively low, but yet is higher than that of the much larger tapir, another favorite food animal. It is rare for the meat of either species to be totally taboo (pp. 83–86 infra). The white-lipped peccary has the larger cruising radius and is less predictable than the smaller species; on the other hand, it is found in larger bands, a circumstance that facilitates multiple kills. It has also been argued (notably in the case of the Yạnomamö) that endemic warfare between large, widely spaced villages in effect creates extensive "buffer-zones," which in turn "constitute reserves [where] many species (especially the larger mammals) may rebound in numbers after a decline in human predation pressures."[101]

---

[94] Campos, 1977: p. 59 (white-lipped peccary in first place, collared peccary in second: note, 22,137 kilograms in error for 2,137 kilograms). See also Smith, 1976: p. 456 (three agrovilas along Brazil's transamazon highway: at one, white-lipped peccary in first place, collared peccary in fourth; at another, white-lipped peccary in second place, collared peccary in fourth; and at the third, longest occupied and now with little virgin forest in close proximity, collared peccary in fifth place, white-lipped peccary not among the thirteen species listed).

[95] Hawkes, Hill, and O'Connell, 1982: p. 386 (white-lipped peccary in first place and collared peccary in sixth in terms of "average calories per consumer day"; but collared peccary and deer in first place and white-lipped peccary in seventh in terms of "rates of caloric resources to handling time" [cost/benefit analysis]).

[96] Lenselink, 1972: p. 40.

[97] Pierret and Dourojeanni, 1967: pp. 10, 15.

[98] Ibid., 1966: p. 273.

[99] Kiltie, 1980: p. 541 (quoting Siskind, 1973: ? p. 204: seven collared peccaries, one white-lipped peccary).

[100] Carneiro, 1974: p. 124 (collared peccary in fifth place, white-lipped peccary not mentioned—below eleventh place). Analysis of faunal remains from the archaeological site of Cerro Brujo (Panama) indicated that the collared peccary was the most important game animal in terms of usable meat (Linares and White, 1980: p. 183).

[101] Ross in Lizot, 1979: p. 153. See also Ross, 1978: pp. 6–8. T. Myers (1976: pp. 354–355) discusses the existence of no-man's lands (based on warfare) and the enhanced opportunities for hunting in such territories, with reference to Amazonia and particularly to the Jívaro (Harner, 1972: p. 56). On the other hand, according to J. Lizot (1977: pp. 513, 515) and N. Chagnon and R. Hames (1979: pp. 910–913) shortage of protein and the need to expand hunting territory cannot account for patterns of warfare among the Yạnomamö.

The collared peccary provides up to 40 pounds of meat, the white-lipped peccary 50 pounds.[102] If the site of the kill is some distance from the camp or village, slaughtered animals may be butchered immediately and the quarters roasted.[103] Oviedo (1526) remarked that in the climate of Tierra Firme (Panama) "fish and meat soon spoil if they are not roasted on the same day that they are killed or caught."[104] The most common method of preserving peccary meat is by smoke-drying. This "[the Indians of Darién] do abroad if they kill a great many *pecary, birds* . . . and bring the pieces home ready dried," wrote Lionel Wafer (1680–1688).[105] Preliminary smoking in the bush may be followed by a similar operation, and more rarely by further processing, in the village.[106] Hans Staden (1547–1555) described from eastern Brazil how dried fish and meat were pounded, passed through a sieve, and reduced to a powder that "last[ed] a long time; for they have not the custom of salting fish and meat."[107] Smoke-drying of peccary meat has been specifically reported from the Cayapa (Ecuador),[108] the Yagua (Peru),[109] the Nambicuara (Brazil),[110] the Canelo (Ecuador),[111] the Chocó (Colombia),[112] the Makú (Colombia),[113] the Tukano (Colombia),[114] the Waiwai (Guiana-Brazil),[115] and the Maya (Mexico);[116] also from Guiana,[117] Panama and Darién,[118] and the lower Ucayalí (Peru).[119] Sun-drying[120] may be associated with salting, but the latter is apparently rare.[121] Until very recently, some of the Northern Kayapó (Brazil) were unfamiliar with both salting and drying.[122] The Bayano Cuna (Panama) either smoke or parboil surplus

---

W. Vickers (1979: p. 20) found no critical shortage of animal protein among the Siona-Secoya of eastern Ecuador.

[102] Nietschmann, 1973: p. 165.
[103] Holmberg, 1950: p. 25 (the Sirionó).
[104] Fernández de Oviedo y Valdés, 1959a: p. 29.
[105] Wafer, 1934: p. 103. Also, apparently, the Makú of Colombia (Silverman-Cope, 1973: p. 74, white-lipped peccary).
[106] Bennett, 1968b: p. 43.
[107] Staden, 1874: p. 132.
[108] Barrett, 1925: p. 75.
[109] Fejos, 1943: p. 41.
[110] Oberg, 1953: p. 89.
[111] Whitten, 1976: p. 178.
[112] Wassén, 1935: p. 86.
[113] Silverman-Cope, 1973: p. 74.
[114] Fulop, 1954: p. 101.
[115] Yde, 1965: p. 123.
[116] Gann, 1918: p. 21.
[117] Bernau, 1847: p. 41.
[118] Cullen, 1866: p. 265; Lothrop, 1948: p. 253.
[119] Pierret and Dourojeanni, 1967: p. 154.
[120] Castelnau, 1850: 2: p. 48 (Rio Tocantins, central Brazil); sun-drying is also mentioned by Wassén, 1935: p. 86 (the Chocó).
[121] André, 1904: p. 134 (Guiana); Farabee, 1968: p. 42 (the Wapishana). Cf. Bernau, 1847: p. 41. Settlers along Brazil's transamazon highway salt or sun-dry game meat (Smith, 1976: p. 459).
[122] J. B. Turner, 1967: pp. 111–112.

meat.[123] The efficacy of these methods depends on the time and care expended; "dried meat," under tropical conditions, is said to keep for anything from a few days to several weeks.[124] At the same time, large quantities of fresh meat may be consumed in feasting or, more often, distributed among kinsfolk and other members of the community. The regular disposal of peccary meat, according to some system of reciprocity, has been widely reported.[125] Some may also be traded. The Makú supply the Tukano with forest products, particularly smoked meat.[126] A document of 1573–1575 refers to the exchange of fish for gold, deer meat, and "barbecued" peccary between Indians along the shores of Lake Maracaibo and others further inland.[127]

Minor products

Hides. Local interest in peccary hides is rather infrequently reported. The Jívaro (Ecuador) are said to value the skins,[128] and the neighboring Canelo use them to cover drums.[129] The latter custom was also reported from sixteenth-century Vera Paz (Guatemala).[130] In the middle of the eighteenth century Abipón women (Paraguay) stitched together skins to make "travelling dresses."[131] A. Métraux refers to large, peccary-skin bags, and maintained that the Chaco Indians "employ skins to a far greater extent than do most South American tribes."[132] Among the Waiwai (Brazil-Guiana), "the cylindrical cover for the quiver containing poisoned arrow points is made from the skin of the *poinko* [white-lipped peccary]."[133]

The Mundurucú "sell hundreds of peccary hides every year to traders,"[134] and hides are occasionally traded by the Tenetehara (Brazil)[135]

---

[123] Bennett, 1962: p. 47.

[124] Holmberg, 1950: p. 34 (two to three days); Neal et al., 1964: p. 58 (a few days); Emst, 1966: p. 61 (weeks, even months); Bennett, 1968b: p. 43 (two weeks or more); Vickers, 1976: p. 101 (a week or so); Ross, 1978: p. 10 (as few as three days).

[125] Stirling, 1938: p. 105 (Jívaro, Ecuador); Fejos, 1943: p. 41 (Yagua, Peru); Oberg, 1953: p. 89 (Nambicuara, Brazil); Henry (1941), 1964: p. 100 (Caingang, Brazil); J. B. Turner, 1967: pp. 105–106, 111–112 (Northern Kayapó); Frikel, 1968: p. 94 (Xikrín [Northern Kayapó]); K. I. Taylor, 1972: pp. 31, 50–58 (Sanumá [Yąnomamö]); Murphy and Murphy, 1974: p. 63 (Mundurucú, Brazil); Langdon, 1975: p. 18 (Barasana and Taiwano, Colombia); Kensinger, 1975: pp. 25, 27 (Cashinahua, Peru); Kaplan, 1975: p. 38 (Piaroa, Venezuela); Dumont, 1976: p. 62 (Panare, Venezuela); Wagley, 1977: pp. 62, 67 (Tapirapé, Brazil). Among some tribes of eastern and southern Brazil (Puri, Caingang, Botocudo, Crahô), the hunter does not eat the meat of any animal that he has killed (Baldus, 1955: p. 195; Henry [1941], 1964: p. 100). This was also formerly the custom of the Sirionó of eastern Bolivia (Holmberg, 1950: p. 32).

[126] Silverman-Cope, 1973: p. 97.

[127] Translated in Breton, 1921: p. 11.

[128] Harner, 1972: p. 56.

[129] Whitten, 1976: p. 43.

[130] Herrera (ca. 1600), 1934–1956: 9: p. 335 (*atambores* [*tambores*]).

[131] Dobrizhoffer [first ed. 1784], 1822: 1: p. 271.

[132] Métraux, 1946a: pp. 284, 292.

[133] Yde, 1965: p. 123.

[134] Murphy, 1958: p. 15.

[135] Wagley and Galvão, 1948a: p. 139 ("Peccary hides bring specially good prices at Neo-Brazilian villages."), 1949: p. 56 (collared and white-lipped peccary).

FIG. 9. A slaughtered peccary and Botocudo Indians, Brazil. J. M. Rugendas, 1835: pl. 1.

and the Sharanahua (Peru).[136] The same has also been reported of the Indians of the Gran Chaco[137] and of *mestizos* settled along the lower Ucayalí[138] and the Río Pachitea (Peru).[139]

---

[136] Siskind, 1973: pp. 70, 169–171 (the skin of the collared peccary being the more valuable, fetching 20 *soles* as against 5 *soles* for that of the white-lipped peccary).

[137] Belaieff, 1946: p. 376.

[138] Pierret and Dourojeanni, 1967: p. 16 (skin of the collared peccary, 22 *soles*, that of the white-lipped peccary, 10 *soles*).

[139] Pierret and Dourojeanni, 1966: p. 273 (collared peccary, 23 *soles*, white-lipped peccary, 9 *soles*, at Pucallpa).

A local market in peccary hides is probably both old and geographically commonplace. They were occasionally used as boot leather and in making chairs.[140] From the second half of the nineteenth century a substantial international demand developed, particularly for the thin, strong, and bristle-marked skins of the collared peccary, used in the manufacture of gloves and jackets. The chief sources of supply were (and remain) northern Argentina and Mexico and the Southwest of the United States. About 1885, San Antonio (Texas) was a collecting point, one firm handling 30,000 in a season.[141] W. J. Hamilton (1939) found that "so great has the demand become, even though the hides bring only 25 or 50 cents, that the animals [collared peccary] have been virtually extirpated over much of their range in the past five years. More than 85,000 Mexican hides came into Nogales, Arizona, a few years ago."[142] Recently, imports to the United States amounted to 18,000 per annum.[143] In 1977 some 3,000 peccary hides were marketed in the province of Salta, Argentina; and between 1972 and 1979 over 300,000 from the whole of Argentina were exported.[144]

Teeth and bone. Peccary teeth (some worked) have been found in archaeological deposits at Cozumel (Preclassic), Dzibilchaltun (Preclassic and Classic), and Mayapan (Postclassic), Yucatán.[145] Recent ethnographic accounts, notably from northeast South America (map 13), refer to their use as cutting implements and as decoration. During initiation rites, Warrau boys (Venezuela) "slashed their chests and arms with peccary tusks";[146] similarly, Betoi hunters (Colombia) made scarification awls of peccary bone (? tusks).[147] The Macusi (Guiana-Brazil) have "knives of peccary teeth,"[148] and among the neighboring Waiwai "the canine teeth of the *pakri* [collared peccary] are inserted in the double scraper, *pákriyórü* (*yórü*, tooth), used for smoothing the bow staff."[149] The Yąnomamö (Venezuela-Brazil) and the Pacaguará (Bolivia) employ the jawbone for the same purpose.[150] Peccary tusks, set in wax, serve as sights on the blowguns of Indians in the Guianas and along the rivers Orinoco and Vaupés.[151]

---

[140] Azara (1801), 1838: 1: p. 119; Villa, 1948: p. 489.
[141] J. A. Allen, 1896: p. 54. In Zavala county in 1886, hides are said to have been used as "currency."
[142] W. J. Hamilton, 1939: p. 368.
[143] C. A. Hill, 1966: p. 7. See also Dalquest, 1953: p. 208 (San Luis Potosí, Mexico); Lewis, 1970: p. 44.
[144] Mares, Ojeda, and Kosco, 1981: p. 199.
[145] Pollock and Ray, 1957: p. 642; Pollock, Roys and Proskouriakov, 1962: p. 377; Hamblin, 1980: p. 241, 1984: pp. 134–135; Wing and Steadman, 1980: pp. 326–327.
[146] Métraux, 1949b: p. 376.
[147] Hernández de Alba, 1948c: p. 398; Métraux, 1949c: p. 581 (quoting J. Rivero, 1736 [1883]).
[148] Farabee, 1924: p. 52.
[149] Yde, 1965: p. 123.
[150] Métraux, 1948a: p. 452; Smole, 1976: p. 185.
[151] Métraux, 1949a: p. 250.

MAP 13. Peccary tusks and bone.

"Necklaces" or "collars" of (worked) peccary teeth have been reported from the Yuruna (Brazil),[152] the Cashinahua (Peru),[153] the Piojé (Ecuador),[154] the Arecuna-Taulipang (Guiana),[155] and the Lacandon (Chiapas, Mexico).[156] Teeth are also used as "children's ornaments" (the Parintintin,

---

[152] Nimuendajú, 1948a: p. 229.
[153] Kensinger, 1975: pp. 170, 174.
[154] Simson, 1886: p. 195.
[155] Schomburgk, 1841: p. 204. For Brazil-Guiana, see also Schomburgk (1840–1844), 1923: 1: p. 274 (vicinity of the River Rupununi, men with big necklaces, *poeng-kere,* of peccary teeth); Burton in Staden, 1874: p. 160 (eastern Brazil, chiefs with necklaces).
[156] Palacios, 1928: p. 150.

Brazil)[157] and for "decorating headdresses, cloaks, drums and shoulder slings" (the Canelo, Ecuador).[158] The Wapishana and the Atorais (Brazil-Guiana) carry peccary tusks as charms or talismans to ensure success in the chase.[159] Lower jaws are prized as hunting trophies by the Waiwai;[160] and among the Canelo lower jaws of both species of peccary are placed in the walls of the *ichilla huasi*, the women's part of the house.[161] Peccary hoofs are made into "bells" (Guiana, and the Northern Kayapó, Brazil)[162] and into rattles worn on the ankles (Tapirapé, Brazil),[163] the wrists (Mbayá, Paraguay-Brazil),[164] or on belts (Yaqui, Mexico).[165]

Bristles. There are a few references to the use of peccary bristles. In bundles they may serve as both brush and comb (Chaco,[166] specifically the Abipón[167]). Sirionó boys (eastern Bolivia) sometimes glue the bristles to their hair.[168] The Tapirapé wear masks decorated with tufts of peccary hair,[169] and C. Nimuendajú observed that among the Parintintin (Brazil) "the hafted ends of [arrow] points now and then have a beautiful fabric of black and white hairs of the [collared] peccary."[170]

---

[157] Nimuendajú, 1948b: p. 287.
[158] Whitten, 1976: p. 171.
[159] Coudreau, 1886–1887: 2: p. 315.
[160] Fock, 1963: p. 159; Yde, 1965: p. 123.
[161] Whitten, 1976: p. 68.
[162] Roth, 1924: p. 465; Frikel, 1968: p. 105.
[163] Wagley, 1977: p. 62.
[164] Métraux, 1946a: p. 336.
[165] Spicer, 1980: p. 12.
[166] Métraux, 1946a: p. 280.
[167] Dobrizhoffer (ca. 1750), 1822: 1: p. 271.
[168] Holmberg, 1950: p. 19.
[169] Wagley, 1977: p. 62.
[170] Nimuendajú, 1948b: p. 289.

E: TABOO, CEREMONY AND MYTH

Rejection of peccary meat (map 14)

Where the peccary is not hunted for food, hunting generally is unimportant (but game not necessarily scarce) and fishing correspondingly important. When first contacted, the Sae (southern Colombia) "ate no meat whatsoever."[1] Other tribes that do not consume peccary meat (Camayurá, Trumaí, Guicuru [Kuikurus]) cluster around the headwaters of the Rio Xingu.[2] The Guicuru kill only to protect their crops and for sport. Dependence on fish is likewise characteristic of the neighboring Apalakiri (Kalapalo), among whom, however, the rejection of peccary (and most other) meat has been described as a "self-imposed taboo."[3] The "traditional aversion" of the Warrau (Orinoco delta) to hunting[4] was also accompanied by a strong interest in fishing. Taboos are difficult to distinguish from indifference (and unfamiliarity) where some alternative resource is valued more highly.

Rejection of peccary meat may reflect the fear that whoever consumes the meat will assume one or another of the characteristics of the animal. The Mataco (Argentine Chaco) "never eat peccary lest they get toothache and their teeth chatter as do those of this animal when it is roused."[5] In the middle of the seventeenth century, the Caribs of Tobago were said to avoid peccary (? and pig) meat in case "they should have small eyes [considered a deformity] like those of that beast."[6] According to A. Simson (1886), the Záparo of the Río Napo do not eat "heavy meats" (peccary, tapir) that "would impede their agility and unfit them for the chase."[7]

Other kinds of calamity may also be anticipated. In the early 1830s, both male and female Yuracaré (eastern Bolivia) refrained from eating peccary meat while clearing forest, so to avoid the risk of being crushed

---

[1] Kirchhoff, 1948a: p. 386.
[2] Oberg, 1953: p. 29; Murphy and Quain, 1955: p. 29; Carneiro and Dole, 1956–1957: p. 181, 1968: p. 245; Villas Boas, 1974: p. 265. Cf. Menget, 1981: p. 12 (the Txicáo).
[3] Basso, 1973: pp. 14, 37, 39.
[4] Wilbert, 1972: p. 89 (the peccary, *báquiro*, is now hunted).
[5] Métraux, 1946a: p. 261. According to Nino (1913: p. 97) the Guisnay, neighbors of the Mataco (and sometimes included with them), hunt the peccary.
[6] Rochefort and Poincy, 1658: pp. 122, 410; 1966: pp. 70, 273.
[7] Simson, 1886: p. 168.

MAP 14. Rejection of peccary meat and restraints on over-hunting.

by falling trees.[8] The Toba (northern Argentina) "fear that the meat of
the collared peccary and domesticated pig will give them ulcers of the
nose."[9] Peccary liver is rejected by the young men of the Guayakí
(Paraguay); to do otherwise would, it is thought, make them poor

---

[8] Orbigny (1826–1833), 1835–1847: 3: pp. 203–204 ("Lorsqu'ils vont abattre les arbres
pour défricher un champ, ils se gardent bien, ainsi que leurs femmes, de manger la chair
du pécari [sanglier de ces contrées], dans la crainte de se voir écraser par les arbres qui
tombent").
[9] Métraux, 1946a: p. 261. Karsten (1932: pp. 38–39) states that the Toba hunt both
species of peccary, but whether for food is not clear.

marksmen.[10] The Siona-Secoya (eastern Ecuador) believe that a pregnant woman who eats peccary meat will give birth to a child with clubbed feet.[11]

Many taboos are associated with the human life cycle, notably pregnancy and childbirth. Some include all meat (the Desana, Makú, Waiwai, Yąnomamö, Shavante, Tenetehara, Eastern Timbira, Tucuna).[12] Both parents of an unweaned Tenetehara child must abstain from "the meat of macaw, white-lipped peccary and tapir."[13] Among the Yąnomamö the flesh of all large game is denied to pregnant women and their spouses.[14] After certain critical events (birth, adolescent rites, parenthood, illness) in the life of a Barasana, peccary meat is among the last to be sanctioned as food.[15] The parents of a sick Kagwahiv (Tupí) child avoid killing or eating the white-lipped peccary.[16] Kayapó women during the first month of pregnancy are not allowed to eat peccary meat.[17] The Jívaro ban meat to parents of a newborn child and to females at the time of their first menses, and similarly the flesh of the white-lipped peccary (among other animals, but not the collared peccary) to betrothed couples for up to two years.[18] Aspiring shamans also avoid "wild boar"[19] (? white-lipped peccary). After the delivery of a first child, the Camacan require the husband to avoid the meat of the peccary, tapir, and monkey.[20] Peccary, deer, and tapir ("strong meat") are included in a similar prohibition by the Cashinahua.[21] Among the Waiwai "a menstruating woman must never eat the flesh of game that has been hunted or caught by dogs, which particularly refers to the *poinko* [white-lipped peccary]. . . ."[22]

In the mind of the Desana (Eastern Tukano), a "young man does not form part of the circuit [of sexual energy] because he is not married; in eating [peccary, tapir, deer, and monkey] useless energy would be accumulated, diminishing the total potential without being able to replace it."[23] At the same time and somewhat paradoxically, a married man, wishing any son to be "spiritually" similar to him, "seeks to

---

[10] Cadogan and Colleville, 1963: p. 442.
[11] Vickers, 1976: p. 213.
[12] Silverman-Cope, 1973: p. 280 (Makú); Macdonald, 1977: pp. 738–739.
[13] Wagley and Galvão, 1948a: p. 142.
[14] Wilbert, 1972: p. 49. Cf. Smole, 1976: p. 251 n. 17 (father of a new-born child denied the meat of tapir, white-lipped peccary, and monkey).
[15] Langdon, 1975: pp. 62 ff.
[16] Kracke, 1981: pp. 107, 136. To kill a tapir or white-lipped peccary is thought to be harmful to a sick person; to eat the flesh, only mildly so.
[17] J. B. Turner, 1967: pp. 142 and Appendix p. v (not nine months, as given in Macdonald, 1977: p. 738).
[18] Karsten, 1920: p. 12; 1935: pp. 116–117, 229, 236–237.
[19] Karsten, 1935: p. 401. Likewise the Siona-Secoya of eastern Ecuador (Vickers, 1976: p. 160, all large game).
[20] Métraux and Nimuendajú, 1946: p. 549.
[21] Kensinger, 1981: p. 161 (for several weeks after the birth of a child).
[22] Fock, 1963: p. 159 (see also p. 158, meat of the *poinko* taboo for two years after the first menses).
[23] Reichel-Dolmatoff, 1971: p. 237.

increase his [sexual] energy by means of restrictions and dietary prescriptions. He should not eat tapir, peccary or monkey meat because the flesh of these animals is impure. . . ."[24] The Desana live along the tributaries of the Río Vaupés, where fish are not abundant. A. D. Wallace (1853) found that some of the Indians of this region ate the meat of the collared peccary, but not that of the white-lipped peccary or the tapir.[25] In general contrast to these customs, pregnant women of the Sirionó (eastern Bolivia) believe that to eat peccary meat will ensure the birth of a valiant and industrious child.[26]

Some Arawak communities of the Pomeroon coast of Guiana refuse the meat of peccaries (and of other animals) found to contain young.[27] The Miskito of Tasbapauni (Nicaragua) reject the collared but not the white-lipped peccary.[28] No other report from Central America has been found. A *relación* of Vera Cruz (1571) states that the flesh of *puercos javalies* was not eaten,[29] but whether this applied to both species and to the Indian or the European population is not clear.

Restraints on overhunting (map 14)

Living creatures are incorporated in many magico-religious beliefs. The idea that animals in general and certain species in particular have "spirit guardians" or "masters" that prevent over-hunting, or may take revenge in the event of over-hunting, is widespread in South and lower Central America.[30] To maintain good relations with the master, and to ensure a sufficient supply of game, the hunter must at all times exercise moderation.

The Cashinahua (Peru) are expected to shoot no more game than the community can reasonably consume.[31] Among the Tenetehara (Brazil), peccaries especially are thought to be "owned" by *Marana ýwa*, Lord of the Forest, and unnecessary slaughter, for example when large herds are trapped by flood-water, is strongly condemned.[32] *Marana ýwa* may punish by inducing sickness or by bringing bad luck in hunting. The spirit-master among some other Tupian tribes is called *Korupira* (*Kuri-Pira*);[33] as *Caapora* the name is also reported from the Shucuru' (eastern Brazil).[34]

---

[24] Ibid., p. 61. Cf. ibid., 1976: p. 313 ("[A] man whose wife is expecting a child should eat neither tapir, peccary nor monkey meat because this might affect the good health of his yet unborn offspring").

[25] Wallace, 1853: p. 485. See also Whiffen, 1915: p. 145 ("Peccary is taboo among many tribes").

[26] Holmberg, 1948: p. 459; 1950: p. 66.

[27] Roth, 1915: p. 297.

[28] Nietschmann, 1972a: p. 54; 1973: p. 167; pork is rarely eaten.

[29] Paso y Troncoso (ed.), 1905b: p. 199.

[30] Zerries, 1954: pp. 93–133 ("Die Herren der Tierarten"); ibid. (1961), 1968: pp. 258–276.

[31] Kensinger, 1981: p. 163.

[32] Wagley and Galvão, 1948a: p. 145; 1949: pp. 58, 103.

[33] Barbosa Rodrigues, 1890: pp. 3 ff.; Nimuendajú, 1915: pp. 290–291.

[34] Hohenthal, 1954: p. 113.

The corresponding figure in Chiriguano mythology is known as *Coquena*.[35]
The Chiripá and Mbyá (Paraguay-Brazil) honor the "Owner of the Pigs"
in ritual song and dance.[36] The Shipaya (Brazil)[37] and the Taulipang-
Arecuna (Guiana-Brazil)[38] also have "peccary masters." According to the
Campa (Peru), certain game animals, including the peccary, are raised
by benevolent spirits that reside in the mountain ranges.[39] The Jirara and
Airico (Colombia), on the other hand, "believe in an evil spirit which
has charge of peccaries."[40] Women of the Canelo (Ecuador) maintain
contact with *Nunghuí*, wife of the forest soul master, by means of black
"stones" that come from the stomach of the peccary.[41]

The Mundurucú "refrain from taking more game than can be eaten
by the village, for it is considered a grievous offence against the spirit
mothers of the animals to commit slaughter or to kill an animal only for
its hide."[42] The "spirit mothers" of the peccary (*daje si*) and the tapir
(*biu si*) are most commonly invoked to protect and to guarantee increase
in numbers. According to the Maracá, "all species are . . . protected by
a particular guardian (*Yorsathë*), which prevents the hunter killing too
many of the same species within a short timespan."[43] The Opaié-
Shavante (southern Brazil) believe that a peccary herd must never be
entirely destroyed.[44] Caduveo hunters (Paraguay-Brazil), in the course of
their active life, are allowed to kill only a limited number of animals of
each species.[45] *Nggïyúdn* is the name given by the Caingang to the "soul"
or presiding spirit of the natural world. When peccaries are scarce, it is
said that "*Nggïyúdn* has become angry with us and has closed up the
pigs in his corral."[46] Scarcity is thus visualized as a punishment; wanton
killing is a cardinal offense.[47] A Yupa folktale recalls how a band of
hunters was punished by *Karau*, Lord of the Animals, for slaughtering
as many as thirty peccaries.[48] Again, in the cosmology of the Miskito,
the white-lipped peccary has a "keeper" (*Wari Dawan*) that prevents
over-hunting.[49] The Cabécar (Costa Rica) attempt to deceive the protector
by giving animals plant names when a hunt is planned.[50]

---

[35] Nordenskiöld, 1924: p. 33.
[36] Cadogan, 1973: p. 98.
[37] Nimuendajú, 1919–1920: p. 1014.
[38] Koch-Grünberg, 1916–1928: 3: p. 422.
[39] Weiss, 1972: p. 193; cf. Calella, 1940–1941: p. 743 (Siona [Sioní], Ecuador–southwestern Colombia).
[40] Hernández de Alba, 1948c: p. 398.
[41] Whitten, 1976: p. 42.
[42] Murphy and Murphy, 1974: pp. 63, 81–82. See also Murphy, 1958: p. 15.
[43] Ruddle, 1970: p. 59.
[44] Ribeiro, 1951: pp. 132–133 (*queixadas*).
[45] Ibid., 1950: pp. 165–166.
[46] Henry, 1964: p. 157.
[47] Métraux, 1946b: p. 470 (Aweikoma-Caingang).
[48] Wilbert, 1974: p. 116.
[49] Nietschmann, 1973: p. 112. Cf. Conzemius, 1932: p. 79.
[50] Stone, 1962: p. 47.

Among the Desana (E. Tukano) the master and protector of animals is known as *Vaí-mahsë*. Peccaries "move in herds about the hills chasing away intruders. These are sacred places that should be avoided, otherwise *Vaí-mahsë* will be angered and will punish the offender with illness. . . . Only a hunter in a state of ritual purity, aided by the invocations of a payé [shaman], dares to go near a hill. . . ."[51] At the same time, the feared "forest spirit" (*boráro*) can take the form of a peccary or a deer. "Peccaries are the favorite animals of the *boráro*, who gives his cry to frighten the hunter and protect his prey. In itself, it is dangerous to follow a peccary into the forest because it might lead the hunter directly to the *boráro*."[52] "Another mechanism that restricts overhunting is this: According to cosmological myths all game animals are associated with certain constellations, as defined by the Tukano . . . a species can only be hunted after its constellation has risen over the horizon. . . ."[53] The Barasana and Taiwano, neighbors of the Desana, regard certain animals, including the peccary, as "soul takers," *sõri masa*. If one is killed without "permission" from the spirit guardian, then illness and death may strike the hunter's community, and the souls of those who die in turn become *sõri masa*, replacing the lost souls of the slaughtered animals.[54]

According to the Sanumá (Yąnomamö), all animals possess a "spirit" (*uku dubi*), which is released on death and can attack those who have ignored specific food prohibitions.[55] Where the "soul" of the peccary is thought to be essentially the same as that of man, as among the Itonama, the flesh may be banned altogether.[56] Medicine men (*baris*) of the Bororo "reincarnate themselves in the very animals which are most valued as food. All these animals are tabooed as food in their natural condition and require a special ceremony which removes their harmful qualities. . . ."[57]

The overall effect of selective taboos may be small, measured in terms of the number and proportion of animals saved from destruction. Whether they serve as effective agents of conservation is also questionable.[58] On the other hand, restraints involving the supernatural do represent attempts to incorporate vulnerable resources in a balanced and self-sustaining cosmology in which man is an integral part of nature.[59]

---

[51] Reichel-Dolmatoff, 1971: p. 82. Cf. ibid., 1976: p. 315 ("[a shaman] will determine the number of animals to be killed when a herd of peccary is reported . . .").

[52] Reichel-Dolmatoff, 1971: pp. 87–88.

[53] Ibid., 1976: p. 313.

[54] Langdon, 1975: pp. 5, 120, 154–155, 183, 282.

[55] K. I. Taylor, 1972: pp. 31 ff.; cf. ibid., 1981: p. 32 (peccaries).

[56] Karsten, 1926: p. 276.

[57] Ibid., p. 277. See also ibid.: p. 294 ("All animals . . . possess a spirit or soul which in essence is of the same kind as that animating man, and which survives the destruction of the body. All animals have once been men, or all men animals." Similarly Siskind, 1973: p. 153 [the Sharanhua]).

[58] Macdonald, 1977: p. 745; Wagley, 1977: p. 67.

[59] See Reichel-Dolmatoff, 1976: pp. 307–318.

MAP 15. The peccary in ceremony and myth.

Hunting rituals, sacrifice, folklore and myth (map 15)

Hunting is a notoriously unpredictable activity. It is, therefore, not surprising that certain rituals (complementary to the supernatural restraints on overhunting) are observed or performed to promote success. Indians of the Guianas (Carib and Arawak) and the lower Orinoco (Warrau) employ "charms" (*binas*) to entice or attract. The *bina* may be a plant, the leaf of which bears a real or imaginary resemblance to a particular animal, for example, the peccary (the Macusi).[60] The Atorais and the

---

[60] Roth, 1915: p. 282. See also Gillin, 1936: p. 180.

Wapishana (Guiana-Brazil) carry peccary teeth as talismans, both to protect themselves and to facilitate success.[61] The Betoi (Colombia) scarify the right arm with peccary bone.[62] Sirionó boys (eastern Bolivia) sometimes glue the bristles of the peccary and the quills of the porcupine to their own hair "so as to make them good hunters of these animals when they grow up."[63]

According to W. E. Roth (1915), when the Arawak of the Pomeroon coast kill a peccary containing young, the latter are buried under the spot where manioc is grated, in order to attract others[64] (the fact that peccaries are fond of manioc may explain the location). With the same purpose, Indians along the Río Vaupés (Brazil-Colombia) bury the head of a peccary where the band was first encountered.[65] The bones of game animals, or more rarely other parts of the body, may be collected or disposed of in some special way (burned, buried in the forest, thrown in streams), thereby to ensure that others of the same species will return.[66] Shamans of the Waiwai (Brazil-Guiana), equipped with a peccary "claw," are reputed to be able to keep droves of the animal close to the village.[67] The Amahuaca (Peru) occasionally drink the blood of the peccary (and of other animals) "for better luck in hunting."[68]

Litanies that refer to peccaries are sometimes recited or sung prior to hunting (Jívaro, Ecuador;[69] Cashinahua, Peru;[70] Mbyá, Brazil-Paraguay;[71] Piaroa, Venezuela;[72] Maya, Yucatán, Mexico[73]). Shamans of the Siona-Secoya (Ecuador) are thought to have the power to "call" game that are normally corralled by their "keepers."[74] The Miskito and the Sumu (Nicaragua) believe that the "owner" of the peccaries "keeps them shut up at times and does not release them unless the *sukya* practices certain

---

[61] Coudreau, 1886–1887: 2: p. 315. Karsten (1935: p. 167) and Zerries (1968: pp. 261–262) refer to various hunting charms. Silverman-Cope (1973: p. 47) states that the Makú (southeast Colombia), unlike the Desana, make little use of them.
[62] Hernández de Alba, 1948c: p. 398.
[63] Holmberg, 1950: p. 29.
[64] Roth, 1915: p. 284.
[65] Coudreau, 1886–1887: 2: p. 171.
[66] Orbigny (1826–1833), 1835–1847: 3 (i): p. 201 (Yuracaré); Nordenskiöld, 1924: p. 123 (Chimane); Karsten, 1935: pp. 167, 173 (Jívaro and other tribes); Wegner, 1936: p. 232 (Chimane); Métraux, 1948b: pp. 500–502 (Mosetene, Chimane, Yuracaré); Holmberg, 1950: p. 91 (Sirionó). Cf. Civrieux, 1959: p. 112 (Cunuana myth, in which the bones of the peccary are thrown into a celestial spring to bring them back to life).
[67] Yde, 1965: p. 122.
[68] Carneiro, 1974: p. 131 (one of several things a hunter can do to bring success).
[69] Karsten, 1935: p. 171.
[70] Tastevin, 1926: p. 160.
[71] Vellard, 1939: p. 171.
[72] Chaffanjon, 1889: p. 203.
[73] Redfield and Villa Rojas (1934), 1962: pp. 350–351.
[74] Vickers, 1976: p. 121. Cf. Calella, 1941: p. 743 (shamans must "apply" to the spirit, *uatti*, of the peccary if hunting is to be successful).

rites of incantation and makes a small offering."[75] Similarly, according to the Makú (Colombia), each "House of Game" has a master

who rears the game and who controls [their] release. . . . Certain men [shamans] have the power and knowledge to travel to the Game Houses in their spirit forms and bargain with the Master for the special release of game animals; they pay him with tobacco smoke, which is his favoured food. . . . [S]ome very powerful shamans are able to close down a Game House altogether, or else to alter the spatial distribution of the monkey hair network [by which animals come to earth], thus controlling the presence and absence of game in particular parts of the forest.[76]

The Mundurucú (Brazil) have several rituals to invoke success. In one dance, 'peccaries' are pursued by hunters and dogs,[77] and in another children take the part of young peccaries, regarded as a great delicacy.[78] The spirit mother of all game animals, *putch ši*, and the spirits of particular species, including the peccary (*daje ši*), must be constantly propitiated to avoid reprisals for human misdemeanors and to ensure the welfare of the community. This is largely the responsibility of the shamans, some of whom may attempt to place the generic spirit in springs near their villages, where animals will thus congregate and hunting will be good.[79] Dances symbolically associated with the peccary have also been reported from the Cubeo (Colombia)—the licentious *hwananiwa* or "wild peccary"[80]—and from the Taulipang-Arecuna (northern Brazil).[81] The Tapirapé (Brazil)[82] and the Guayakí (Brazil-Paraguay)[83] celebrate a successful peccary hunt in song or dance.

Sacrifice and religion. Some sacrifices of peccary were apparently part of burial ceremonies (Guatemala;[84] province of Santiago del Estero and the Chaco of Argentina[85]). The ceremonial slaughter of captured peccaries is practiced by the Eastern Timbira (Brazil),[86] the Jívaro (Ecuador),[87] and the Cashibo and Shipibo (eastern Peru).[88] The Maya in early post-

---

[75] Conzemius, 1932: p. 79.
[76] Silverman-Cope, 1973: pp. 254, 275.
[77] Horton, 1948: p. 280.
[78] Farabee, 1917: pp. 135–136. See also Strömer, 1932: pp. 119–120; Murphy, 1958: p. 60 (ceremony to gratify animal spirits, men adorned as peccaries, tapirs, and monkeys).
[79] Murphy, 1958: pp. 13–15.
[80] I. Goldman, 1963: p. 236.
[81] Koch-Grünberg (1911–1913), 1916–1928: 3: pp. 159, 422. See also Brett, 1868: p. 374; Beebe et al., 1917: p. 465; Roth, 1924: p. 479.
[82] Wagley, 1977: p. 62.
[83] Cadogan, 1973: p. 98.
[84] Borghegyi, 1965: p. 23.
[85] Rusconi, 1931b: pp. 228–240 (the extinct genus *Platygonus*).
[86] Nimuendajú, 1946: p. 44.
[87] Farabee, 1972: p. 121.
[88] Roe, 1982: p. 110.

Conquest times sacrificed peccaries.[89] This has also been inferred from faunal remains at Mayapan,[90] and from codical representations of peccaries caught in snares.[91] In the cosmology of the Maya, peccary (and deer) were closely associated with agriculture, seasonal change, and annual renewal.[92] The pre-hispanic *Codex Tro-Cortesianus* shows a fertility goddess surrounded by deer and peccary, among other animals (fig. 10). The Pacaguará (Bolivia) "worshipped their deities in the guise of . . . a peccary's or some other animal's head."[93] Pre-Columbian models of the peccary (supra p. 30) may have had similar significance.

Folklore and myth. Peccaries are prominent in Amerindian folklore. Most instructive are accounts of their mythical relationships with man. The Yauavo (Peru) are the "Peccary people" (*Yawabu*).[94] In a folktale common to the Guayakí, the Chiripá, and the Mbyá (Paraguay-Brazil) a young Indian is obliged to marry a peccary.[95] Certain bands of the Mataco, the Toba, and the Chamacoco (Argentina-Paraguay) are named after animals, including the peccary.[96] The *yehsë* is regarded as the progenitor of one of the sibs (*Yehsë-porá*, "Sons of the peccary") of the Desana (Colombia).[97] A number of widely distributed myths relate how, in a variety of circumstances, people (men, women, and children) were transformed into peccaries (the Warrau, Venezuela;[98] the Sharanahua, Peru;[99] the Cariri, Brazil;[100] the Crahõ and the Kayapó, Brazil;[101] the Tenetehara, Brazil;[102] the Mundurucú, Brazil;[103] the Bororo, Brazil-Bolivia;[104] the Mataco, Argentina).[105] To shed light on the role of the peccary as an "intermediary" between animal (jaguar) and man (exemplified in tales belonging to the Kayapó, Opaié-Shavante, and Tucuna, Brazil), Lévi-Strauss analyzed versions of the transformation myths of the Kayapó, Tenetehara, and Mundurucú,[106] and concluded:

[89] Tozzer (ed.) in Landa, 1941: pp. 5 n. 24, 115 n. 528; Pohl and Feldman, 1982: p. 295.
[90] Pollock and Ray, 1957: p. 640.
[91] Hamblin, 1980: p. 245.
[92] Thompson, 1970: p. 370; Pohl and Feldman, 1982: pp. 296, 299.
[93] Métraux, 1948a: p. 452.
[94] Ibid., 1948c: p. 660. See also Reichel-Dolmatoff, 1975: p. 58 ("women of peccary group," among the Kogi, Colombia).
[95] Cadogan, 1973: p. 98.
[96] Métraux, 1946a: p. 302.
[97] Reichel-Dolmatoff, 1971: p. 200.
[98] Wilbert, 1970: nos. 42, 69, 76, 95, 205.
[99] Siskind, 1973: p. 153.
[100] Lowie, 1946c: p. 559.
[101] Métraux, 1960: pp. 28–29; Wilbert, 1978: nos. 24, 32, 62, 64, 95, 96.
[102] Wagley and Galvão, 1949: p. 134.
[103] Murphy, 1958: pp. 70–73.
[104] Lévi-Strauss, 1969: pp. 94–95.
[105] Métraux, 1939: p. 61.
[106] Lévi-Strauss, 1969: pp. 67–68, 82–87.

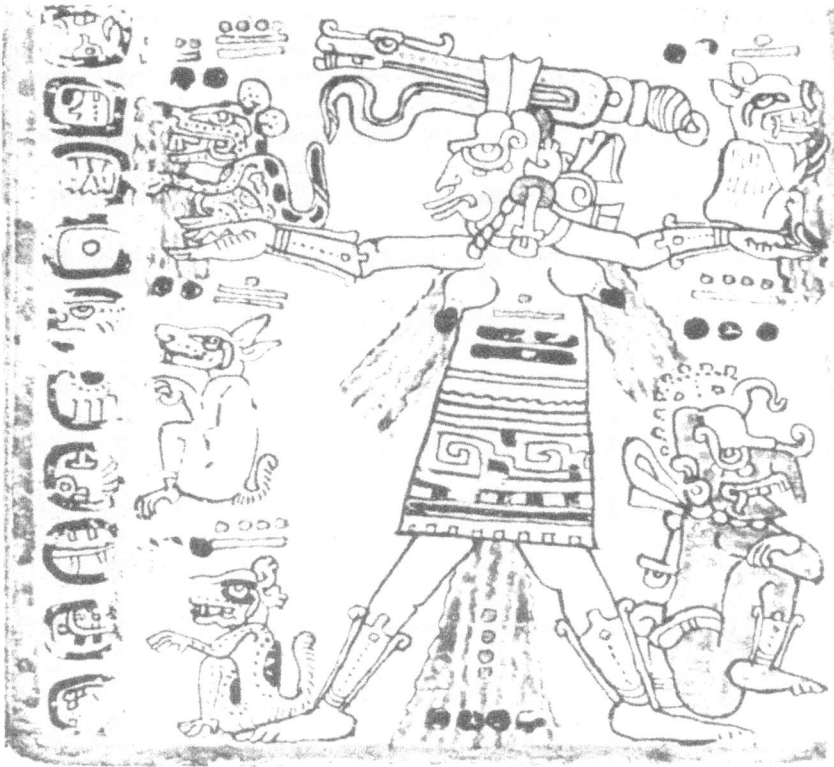

FIG. 10. Fertility goddess with a peccary (upper right) and other animals. *Codex Tro-Cortesianus* [*Madrid*] (Mayan, Pre-hispanic), A. Anders, 1967: p. 30 b.

The *caititus* [collared peccary] and peccaries [white-lipped peccary] are therefore semi-human: the former synchronically, since they constitute the animal half of a pair whose other member is human; and the latter diachronically, since they were human beings before they changed into animals.

If, as may be the case, the Mundurucú and Cayapo myths [involving the white-lipped peccary] preserved the memory of a technique of hunting that was no longer practised and consisted of driving peccaries into enclosures where they were kept and fed before being killed according to need, the first contrast is reduplicated by the second: semihuman on the mythic level, the peccaries could be semidomesticated on the level of techno-economic activity. If so, it would have to be admitted that the second aspect explains, and is the basis of, the first.

This may be usefully compared with the observations of G. Reichel-Dolmatoff:[107]

---

[107] Personal communication (unpublished manuscript: pp. 40–42).

In the view of most Vaupés Indians [Northwest Amazon], peccary are rather disgusting creatures [although their meat is highly appreciated] . . . foul-smelling, always foraging and grunting, and openly promiscuous. . . . All these character-istics make them readily comparable to the Makú Indians, especially to Makú women. The Desana, Pira-Tapuya and Tukano will often speak of these similarities and in many myths and tales Makú–peccary comparisons are described in detail. In many ways the Makú are still not thought to be "quite human"; thus they constitute a continuum between nature and culture and this is expressed in [the] peccary's occasional forays into cultivated fields . . . the Desana say that women who, in their animal image, eat forest products, as tapir and deer do, are outsiders, but that those like the peccary, who steal and eat cultivated field fruits, are "sisters," that is, they can be married.[108]

---

[108] On the inferior position of the Makú, see also P. H. Allen, 1947: p. 569; Silverman-Cope, 1973: pp. 99, 195.

F: TOWARDS DOMESTICATION

Before the arrival of Europeans, the Muscovy duck, the turkey, and probably the (introduced) guinea pig were the only fully domestic animals within that part of the New World occupied by the two peccaries. On the other hand, many tame animals were kept, especially in the humid tropics of South and Central America. Captured peccaries are common enough, doubtless far more so than the adventitious evidence plotted in Map 16 (which does not include regional observations) would indicate.[1] The collared species is most often mentioned, but white-lippeds are certainly also reared.[2] The latter, apart from the occasional tapir (*Tapirus* spp.),[3] are the largest of the semi-domesticates.

Early evidence (prehistoric to ca. 1625) of the keeping of peccaries in the Mayan lowlands and Central Mexico, and in lower Central America and northern Colombia, has already been presented (pp. 38–39).[4] Chiefly remarkable are the records of trade in reared peccaries in the Isthmus and the region around the Gulf of Urabá (the distribution is similar to that wherein nets were used to capture game, Map 11). Such trade has apparently continued to the present day in some remote areas. "A favorite commodity for exchange, especially with the Cabécares [Talamancan region, Costa Rica], is the wild pig. . . . These [tamed] pigs are fattened for a given feast . . . or are interchanged for salt and woven mantas with the Borucas."[5] The sale of peccaries has also been reported

---

[1] Tribal references, other than those referred to below, in Boman, 1908: 1: pp. 90–91 (? Diaguites, Argentina); I. Goldman, 1948: p. 772 (Vaupés-Caquetá region); Métraux, 1948e: p. 101 (Tupinamba, Brazil); Barker, 1953: p. 443 (Guaika, Venezuela).

[2] Kerr ([Gmelin]Linnaeus), 1792: p. 352; Jardine, 1843: 9: p. 236; Buffon (ca. 1780), 1884: 9: p. 234; Schomburgk (1840–1844), 1923: 2: pp. 128–130, Farabee, 1924: p. 33 (Macusi, Guiana-Brazil); Miller, 1930: p. 19 (southern Mato Grosso, Brazil); Nimuendajú, 1946: p. 75 (Eastern Timbira, Brazil); J. B. Turner, 1967: p. 138 (N. Kayapó, Brazil); Nietschmann, 1973: p. 150 (Miskito, Nicaragua). Husson (*Mammals of Suriname*, 1978: p. 356) states that the young of *D. pecari* "cannot be tamed," but this is clearly not generally true. Richard Burton (in Staden, 1874: p. 160 n.) remarked that the collared peccary was "hard to tame." He may have been referring to adult specimens; the evidence for juveniles is otherwise.

[3] Sack (*Surinam*, 1805–1807), 1810: p. 241 ("they become very tame. . . . I never could learn whether a trial has been made to breed them when domesticated."); Zevallos (1610), 1886: p. 157 (*dantas mansas*, Panama); Woodroffe, 1914: p. 63 (upper Amazon); Kracke, 1981: p. 105.

[4] For the eighteenth century, see: Stedman (1772–1777), 1796: 1: pp. 355–357 (Surinam); Dueñas (1792) in Izaguirre Ispizua, 1922–1929: 8: p. 245 (Pano, eastern Peru); Caulín (1779), 1966: 1: p. 73 (Nueva Andalucía); and cf. Figueroa (ca. 1650), 1904: p. 207 (Maina); Ximénez (1722), 1967: p. 57 (Guatemala).

[5] Stone, 1962: p. 41. See also Frost, 1974: p. 160.

MAP 16. Tame or penned peccaries and trade in peccaries.

from Surinam.[6] In general, however, the traffic has probably declined with the introduction of the domestic and more prolific European pig.

Peccaries encountered in villages have often been described as "pets," being cared for (food and protection, particularly from dogs) by women and children,[7] but also free to roam and to root for themselves.[8] In fact,

---

[6] Sack (1805–1807), 1810: pp. 242–243; Kloos, 1971: p. 59.

[7] Dobrizhoffer (1784), 1822: 1: p. 89 (Abipón, Paraguay); Bell, 1899: pp. 174–175 (Miskito Coast); Nordenskiöld, 1912: p. 50 (Chorotí and Ashluslay, Paraguay); Nimuendajú, 1939: p. 95 (Apinayé, Brazil); Vellard, 1939: p. 88 (Guayakí, Paraguay); J. B. Turner, 1967: p. 138 (N. Kayapó, Brazil); Lewis, 1970: pp. 44–45 (Guatemala); Smole, 1976: p. 185 (Yąnomamö, Brazil); Roe, 1982: p. 110 (Shipibo, eastern Peru). According to Holmberg (1950: p. 29), the Sirionó keep young animals as pets, but, since they are not fed, they do not live for long.

[8] Azara (late eighteenth century), 1838: p. 115 (Paraguay); Hatt, 1938: p. 336 (Yucatán,

they do not usually wander far, and apparently never voluntarily return to the forest. They may serve as "watches,"[9] raising the alarm in the event of intruders, like some tame birds; or they may be kept to "attract disease" and thereby protect their owners.[10] The Cainguá (1894) of northern Argentina raised peccaries (and other animals) more as "curiosities," to amuse, than as a potential food supply.[11] Nevertheless such "pets" may eventually be eaten,[12] especially when they get older and become less appealing.

Pets, captured in the wild, are usually kept singly and are subject to imprinting on humans, a circumstance making it impossible or very unlikely that they will reproduce. Tame peccaries sometimes become very attached to their owners.[13] More opportune circumstances exist wherever several individuals are penned, either for purposes of trade or as a general reserve of food,[14] or again to provide for sacrifices, ceremonies and feasts. The latter would appear to be the more powerful motive. The Jívaro (Ecuador) mark the end of a particular celebration by slaughtering "a large number of young peccaries, which have been kept fat for the occasion"; the meat is then distributed to guests for their journey home.[15] A Jívaro myth on the "origin of domestic pigs" relates how a young female peccary was raised and later kept in an enclosure. This proved to be too frail and was invaded by other peccaries.[16] The idea of enclosures or stockades for captured peccaries is also suggested in myths preserved by the Mundurucú (Brazil),[17] the Kayapó (Brazil),[18] the Chiriguano (southern Bolivia),[19] and, apparently less explicitly, by the Tenetehara (Brazil).[20]

Mexico). See also Woodroffe, 1914: p. 71 (pets, including peccaries, "in every cottage": upper Amazon); Farabee, 1918: pp. 39–40 ("grow up around the house": Wapishana, Guiana-Brazil); Rohl, 1959: p. 142 ("ambas especies son comunes en sus hábitos de vida"— Venezuela); Humboldt (1811), 1966: 3: p. 51 (frequently "in the cottages of the natives of South America"), and cf. ibid., 1852–1853: 2: p. 269 (Maipure, Piaroa, Venezuela). Some of these animals may have been penned.

[9] Pennington, 1969: p. 146 n. 57 (Tepehua, Mexico); Pohl, 1976: p. 203 (Maya, Peten, Guatemala).

[10] Reichel-Dolmatoff, 1971: p. 186 (Desana: "forest animals," not specifically the peccary).

[11] Ambrosetti, 1894a: pp. 702, 726.

[12] Gann, 1918: p. 25 (Maya); Vellard, 1939: p. 88 (Guayakí, Paraguay); Guppy, 1958: pp. 234–235; Maybury-Lewis, 1967: p. 37 (Akwĕ-Shavante, Brazil); Kloos, 1971: p. 59 (Surinam). On the other hand, Roth (Guiana, 1924: p. 556) states that pets are not eaten; earlier (p. 555) he refers to peccaries as "pets." Similarly, Smole, 1976: p. 185 (Yanomamö). The Northern Kayapó (J. B. Turner, 1967: pp. 137–138) eat neither pets, including peccary, nor domestic animals (pigs, poultry); the Kagwahiv [Tupí] (Kracke, 1981: p. 105) also reject "pets," which may include all domestic animals.

[13] For an early illustration, see Ximénez (1722), 1967: p. 58.

[14] Gross (1975: p. 533) refers to the keeping of peccaries "for later consumption"; but they are not bred.

[15] Farabee, 1922: p. 121.

[16] Wavrin, 1932: pp. 132–133.

[17] Kruse, 1951: p. 1006; Murphy, 1958: p. 71.

[18] Métraux, 1960: pp. 28–29.

[19] Nordenskiöld, 1924: p. 33.

[20] Wagley and Galvão, 1949: p. 134. See also Barbosa Rodriguez, 1890: p. 48 ("um bom curral").

Among the Eastern Timbira (Brazil), "tame" peccaries

must generally be kept in cages to guard against constant brawls with dogs. The peccaries receive the names of human beings. At the terminal solemnity of certain major festivals a peccary must be ceremonially killed. . . . The inmates of the house loudly bewail the death, while no particular fuss accompanies the slaughtering of a domestic pig.[21]

Tame peccaries are also sacrificed by the Cashibo (eastern Peru) as part of a fertility rite, and by the neighboring Shipibo after a female puberty rite.[22]

According to Pedro Lozano (ca. 1750) the Guayakí had the "foresight" (*providencia*) to "domesticate" (*domesticar*) a few *jabalí*.[23] The lowland Maya occasionally rear peccaries.[24] Penning juveniles is "common practice" among the contemporary Guaymí of Bocas del Toro (Panama).[25] Some Cuna erect pens inside their houses; the peccaries are not tamed or kept as pets, but are reared for food.[26] "The Chocó [of the Pacific coast of Colombia] tether and fatten in the hut young peccaries that they chance to trap alive during a hunt."[27] On the Caribbean slope of the Talamancan highlands (Costa Rica), captured peccaries are sometimes confined to "a semicircular pen of sticks . . . built against the inside wall of [a] dwelling."[28]

Penned animals must of course be regularly fed, a condition that could be a serious disincentive. On the other hand, peccaries are omnivorous. They also breed at any time of the year. If penning were practiced for any considerable length of time, fortuitously adapted pairs, of appropriate age, might be brought together. Since, however, peccary litters are small (usually not more than two), the long-term advantages of breeding in captivity would not be readily apparent. Again, perhaps the most favorable conditions existed where peccaries were regularly traded or were periodically sacrificed. Probably propitious, too, were certain widely held ideas: that peccaries were "semi-domestic"—raiding cultivated crops; that they were in some way related to man, even "semi-human"—transformation myths; and that they were kept in enclosures by spirit masters—"domestic animals of their protectors."[29] Garden hunting, which

---

[21] Nimuendajú, 1946: p. 75; Lowie, 1946b: p. 482. The Guayakí (Paraguay) also use cages or baskets.

[22] Roe, 1982: pp. 110–111. Roe remarks that "peccaries here, as elsewhere in the lowlands, are associated with women" (see supra p. 94).

[23] Lozano, 1873–1874: 1: p. 415.

[24] Nimis, 1982: p. 315. See also Hatt, 1938: p. 336 (Ticul, Yucatán); Steggerda, 1941: p. 145 (collared peccary as a "pet").

[25] Linares and White, 1980: p. 186.

[26] Bennett, 1962: pp. 39–40. Linné (1929: pp. 130–131) observes that the Chocó (northern Colombia) have pig pens under the floor staging of their pile dwellings, and he speculates that peccaries were kept in the same fashion.

[27] West, 1957: p. 245.

[28] Stone, 1962: p. 13.

[29] Redfield, 1942: p. 118.

has been suggested as a possible substitute for domestication,[30] may also have been in the nature of a prelude.

Successful breeding in captivity is unlikely to have escaped attention. Yet the only known allusion to breeding is in Juan de Velasco's *Historia del Reino de Quito* (1789). His annotated list of *especies de puercos* includes *"Huasi-cuchi,* o puerco casero [*huasi* or *wasi,* Quechua "house"], no porque lo sea, sino porque se domestica fácilimente, y procrea con los llevados de otras partes. Es algo más bajo, pero talvez más largo que el europeo, negro con faja blanca que la ciñe todo el cuerpo y de carne muy gustosa."[31] This appears to refer to the collared peccary.

---

[30] Linares, 1976a: p. 331.
[31] Velasco, 1946: 1: p. 118.

G: ANIMAL DOMESTICATION IN THE HUMID TROPICS

Relatively few domestic animals originated in the humid tropics. The peoples of South and Southeast Asia contributed to the domestication of the common fowl (*Gallus gallus*), the duck (*Anas platyrhyncha*), the pig (*Sus* spp.), and perhaps the dog (*Canis familiaris*). The principal herd animals (sheep, goats, cattle, horses, and camels) belong to the periodically arid and more open lands of West and Southwest Asia. In Africa domestic animals, apart from the cat, the ass (Nile valley/Ethiopia), the guinea fowl (? forest/savanna margins) and the Egyptian goose, are everywhere intrusive. The Guinea hog (*Potamochoerus porcus*) of West and Central Africa is reared but, as far as is known, has never been regularly bred. In the New World the forested lowlands provided only the Muscovy duck (*Cairina moschata*). The llama, alpaca, and guinea pig are Andean, the turkey Mexican,[1] and the cultivated cochineal insect[2] one or other or possibly both. The relevant regional similarities are thus:

South and Southeast Asia	West and Central Africa	Lowland tropics of the Americas
pig	bush pig (*reared*)	peccary (*reared*)
common fowl, duck	guinea fowl	Muscovy duck

The bush pig and the peccary were probably on the way to being domesticated at the beginning of the Age of Discoveries. The peoples involved were essentially horticultural (variably combined with hunting, fishing, and collecting) and, to an increasing extent, sedentary (or permanently settled in exceptionally favorable circumstances). From the very remote and largely notional "origin" of farming, the indigenous processes of cultural change among the unspecialized cultivators of the humid tropics appear to have been extremely slow. Influences emanating from "advanced" (culturally more specialized) societies were more powerful in the Old World, with its greater stock of domestic animals, the invention and diffusion of the plow, and the comparatively early emergence of forms of urban/theocratic organization—facilitated by the greater per capita food surplus possible with plow-farming. In the New World all cultivation was horticultural (based on hand implements) and generally unspecialized until the arrival of Europeans.

---

[1] The domestication of birds (Old and New World) and other New World domesticates are considered in Donkin, n.d., which is chiefly devoted to the Muscovy duck.

[2] Donkin, 1977: pp. 14–17, 32–35, 51.

FIG. 11. "Men [Akwĕ-Shavante, Brazil] returning from the hunt with a captured peccary." D. Maybury-Lewis, 1967: pl. 69.

The domesticates of the humid tropics (and the guinea pig) are scavengers; the pigs were also initially crop-robbers. Rather like volunteer plants, they invited attention. The major herd animals, domiciled in the lands of (later) "high" civilization in the Old World and the New, seem to have required more positive action, and thus greater motivation, on the part of man. The keeping of captured juveniles of many species, out of curiosity or for mere pleasure, is indeterminately old and appears to be particularly, but by no means exclusively, associated with the horticultural societies of the tropics. Chance breeding in captivity of a very few (adaptable) species might have led to domestication for primarily economic reasons, provided that supplementary feed presented little or no problem (omnivorous species, including scavengers—again pigs and poultry) and that hunting was relatively unrewarding. An interest in breeding animals (after chance occurrences), initially for ceremonial purposes, may have been common to all areas of successful domestication and a particularly powerful motive where difficult herd animals were involved—Southwest Asia and the Andes, in association with the origins of "organized" religion.

The keeping of pets (usually nurtured by women and children) and the incorporation of animals in folklore, mythology, and religious observances have been widely reported from the humid tropics. The American species include the peccary. But for European intervention, its domestication would probably have been achieved under one or the other, or the convergence, of two sets of circumstances: (i) ceremonial— the rearing of peccaries for purposes of sacrifice on specific occasions, and (ii) economic—trade in fattened animals, locally and very gradually substituting for hunting. Both practices are likely to have been more common before the profound cultural and demographic changes set in motion by the conquest, and also more common than the evidence, from the period of European contact onwards, can possibly prove. The historical and ethnographic record is related to the territorial expansion of exploration, settlement, and scientific enquiry. All that can be inferred is that the peoples of lower Central America and northern South America engaged in activities that could have led to domestication. Here Muscovy ducks were bred, many animals were kept as curiosities, and the peccary was sacrificed, traded, and indeed carried by man to some of the nearer islands. Future research may show, however, that this was only a peripheral manifestation of developments closer to the Amazonian heartland of South America.

APPENDIX
ADDITIONAL VERNACULAR NAMES FOR THE PECCARY

NORTH AND MIDDLE AMERICA	CENTRAL AMERICA	SOUTH AMERICA
Macromayan	Chibchan	Tupí-Guaraní
Other languages	Misumalpan	Cariban
	Other languages	Arawakan
		Tucanoan
		Ge
		Panoan
		Other languages

Tribe	collared peccary	white-lipped peccary	species inadequately identified	Authority
NORTH AND MIDDLE AMERICA				
Macromayan				
Huastec			altéolom, olóm	Tapia Zenteno, 1767 [Vocab.]
Ixil			c'op chal, cop-chan	Stoll, 1884: p. 56 Stoll, 1884: p. 56 Stoll, 1887: p. 111
Mixe	poo'p iitsɨm	iitsɨm, tsapts iitsɨm	bok-yoy, pajpi itz'ami	Lehmann, 1920: 2: p. 774 Schoenhals and Schoenhals, 1965 [Vocab.]
Mixe [Western]			azengók	Beals, 1945b: p. 136
Popoluca			mog uíchim	Clark and Davis de Clark, 1960 [Vocab.]

Tribe	collared peccary	white-lipped peccary	species inadequately identified	Authority
NORTH AND MIDDLE AMERICA (*Continued*)				
Totonac			*qui'hui' pa'xni' qu'ihui p'axni*	Aschmann, 1962: p. 33 Aschmann and Dawson, 1973 [Vocab.]
Totonac [Tajín]			*kiwipašni*	Kelly and Palerm, 1952: p. 74
Other languages				
Cáhita (Mayo/Yaqui) (Mayo)	*húya kówi juya cohui*			Beals, 1945a: p. 10 Collard and Collard, 1962 [Vocab.]
Cora	*yaíjubej*			McMahon, 1959 [Vocab.]
Mayo—see Cáhita				
Nevome (Pima Bajo)	*tasicori, tásikor*			Pennington, 1979–1980: 1: p. 211, 2: p. 67
Popoloca			*cuchingâná, culúchigñá*	León, 1912: p. xix
Seri	*ziix ína quíocs*			Moser and Moser, 1961: p. 159
Tepehuan				Pennington, 1969: p. 70
Yaqui—see Cáhita				
Zapotec	*taísoli*		*becobehueguixi, behuetani, dani pêhue, pêhue-tári, pehue-quijxi bihui gui'xhi' behuetaminatoxco, bihuitaninotoxoo*	Junta Colombina de México, 1893 [Vocab.] Seler, 1909: p. 402 Pickett, 1959: p. 71 Martín del Campo, 1960: p. 68

CENTRAL AMERICA

Chibchan

Boruca		kra-mi-shuk		Gabb, 1876: p. 594
Bri-Bri	kåsir, kåsir-ūiko			Lehmann, 1920: 1: p. 319
Cabécar	kas'-ri, kas-i-ri'			Gabb, 1876: p. 594
Cuna		yánu	upicuácal	Lehmann, 1920: 1: p. 140
				Lehmann, 1920: 1: p. 130;
	huedar			Bennett, 1962: p. 42;
	huédar, guatarra			Mendez, 1970: p. 239
Cuna [Tule; San Blas Cuna]			yanu-chapurri	Bennett, 1962: p. 42
				Mendez, 1970: p. 243
				Cullen, 1868: p. 172;
				Lehmann, 1920: 1: p. 129
Guatuso	áxār	uxúuti		Lehmann, 1920: 1: p. 407
Guaymí [Muci]			béto	Lehmann, 1920: 1: p. 164
Guaymí [Muríre]			tódra	Lehmann, 1920: 1: p. 164
Guaymí [Sabanero]			tódra	Lehmann, 1920: 1: p. 164
Térraba	shtiko			Lehmann, 1920: 1: p. 266
Tiribí	shtuk'-o			Gabb, 1876: p. 594;
	shtiko			Lehmann, 1920: 1: p. 266
	shto-ko			Mendez, 1970: p. 239

Misumalpan

Miskito		uári		Lehmann, 1920: 1: p. 508
Rama			ūlkan, nūlkán	Lehmann, 1920: 1: p. 450
Sumo	mólókōs			Lehmann, 1920: 1: p. 508
Úlua	moolakoos, mūlukús			Lehmann, 1920: 1: pp. 574, 576

Tribe	collared peccary	white-lipped peccary	species inadequately identified	Authority
CENTRAL AMERICA (Continued)				
Other languages				
Chilanga			mapit-cotán, mâpit kôtan	Lehmann, 1920: 2: pp. 694, 716
Jicaque			siba, nam	Von Hagen, 1943: p. 86
Lenca			napir	Lehmann, 1920: 2: p. 675
Pipil (of Izalco)			kúyâmēt	Lehmann, 1920: 2: p. 1047
Pipil (of Guatemala)			cojtan-coyani	Lehmann, 1920: 2: p. 1065
Subtiaba			âga-ramâ	Lehmann, 1920: 2: p. 944
Xinca			cargua-juxo	Brinton, 1884: pp. 93, 95; Lehmann, 1920: 2: p. 733
			âsu	Lehmann, 1920: 2: p. 750
			carau-áso, crauva-áxo	Lehmann, 1920: 2: p. 736
			caraguá árru	Lehmann, 1920: 2: p. 750
SOUTH AMERICA				
Tupí-Guaraní				
Caliana		ãmô		Koch-Grünberg, 1916–1928: 4: p. 315
Mauè			hamaot	Coudreau, 1897b: p. 176
Tapirapé	chiunga			Baldus, 1970: p. 212
Uara Guaçu		tupitono		Martius, 1863: p. 477, 1867: 2: p. 18
Yuruna [Juruna]			fouya	Coudreau, 1897c: p. 176

Cariban

Arara	poon			Coudreau, 1897c: p. 204
Cumanagoto	tirigua	cuacua		Caulin (1779), 1966: 1: p. 73; Civrieux, 1980: p. 164
Hianácoto-Umáua		uóto, wóto, uotóime		Koch-Grünberg, 1908: p. 45
Macusi		whinga		Thurn, 1883: p. 109; Whiffen, 1915: p. 128
Makiritare	kaampo			Chaffanjon, 1889: p. 343
Maracá		kokoona		Ruddle, 1970: p. 43
Motilón		kasáre	guachari	Jahn, 1927: p. 347
[? Mape, Bure, Chaké]				

Arawakan

Arawak	matula		Goeje, 1928: p. 226
Arekena	dohála		Koch-Grünberg, 1911: p. 132
Arua	urumaru		Goeje, 1928: p. 226
Baniwa	soára, tsoára		Koch-Grünberg, 1911: p. 132; Goeje, 1928: p. 226
Baré	arúa		Koch-Grünberg, 1911: p. 132; Goeje, 1928: p. 226
Cariaya			Martius, 1863: p. 477
Carútana	sòara	ayza	Koch-Grünberg, 1911: p. 132; Goeje, 1928: p. 226
Cayuishana	yamughato	nauá	Martius, 1867: 2: p. 259
Culino	uná	iuá:ra	Martius, 1867: 2: p. 244
Guinau	weli:zi	inarra	Koch-Grünberg, 1916–1928: 4: p. 283; Goeje, 1928: p. 226

Tribe	collared peccary	white-lipped peccary	species inadequately identified	Authority
		SOUTH AMERICA (*Continued*)		
Jumana	yamukaische			Martius, 1867: 2: p. 252
Mandauáca	arúa, alúa			Goeje, 1928: p. 226; Koch-Grünberg, 1916–1928: 4: p. 295
Marawa		arua, arûa		Martius, 1867: 2: p. 224; Goeje, 1928: p. 226
Mariate		kâpéna		Martius, 1867: 2: p. 267
Masco			ote	Farabee, 1922: p. 77
Mayopityan		bita		Farabee, 1918: p. 285
Saraveca	imiazare			Goeje, 1928: p. 226
Uainumá		capéna, cabéna		Martius, 1867: 2: p. 248
Wapishana		bitci		Farabee, 1918: p. 221
Waraicu [Uairacu]		alúa		Martius, 1863: p. 477
Yucúna	hapié	heęrú		Koch-Grünberg, 1911: p. 132
Tucanoan				
Bará	yehépuro	yehé		Koch-Grünberg, 1914a: p. 170
Buhágana	yeęsépuro	yeęsé		Koch-Grünberg, 1914a: p. 578
Buhágana [Ömöä]	yesépotiro	yesé		Koch-Grünberg, 1914a: p. 578
Buhágana [Sära]	kîyeęsé	yeęsé		Koch-Grünberg, 1914a: p. 579
Buhágana [Tsöloa]	kîyeęsé, kîyehe	yehe yehe		Koch-Grünberg, 1914A: p. 578

Tribe				Reference
		yexsé		Koch-Grünberg, 1914a: p. 579
Carapaná	kięse motiro	kięseka kapãn		Koch-Grünberg, 1915–1916: p. 430
Cotoxó [Coto]		kuga, hüahiä		Martius, 1867: 2: p. 158
Cubeo	oálihęhę, oálihehę, oalihęhę	oáliobętiue, oáliue, oalí obędiu		Koch-Grünberg, 1915–1916: pp. 126, 430
Cueretu		záesé		Martius, 1863: p. 477, 1867: 2: p. 166
	liámękiatu	sęxsę	tschetshé	Martius, 1867: 2: p. 285
		yexsé		Koch-Grünberg, 1914a: p. 822
Desana	yexsépuru, yexsépulu	yexsé	tshetse	Pfaff, 1890: pp. 603, 605
	yehsë	yehsë		Koch-Grünberg, 1914a: p. 822
				Reichel-Dolmatoff, 1971: p. 200
Erúlia	kíyęxsę	yexsé		Koch-Grünberg, 1914a: p. 579
Omöa—see Buhágana				
Palänoa	ye(x)sépuro, ye(x)sépulo	ye(x)sé, hálayexse		Koch-Grünberg, 1914a: p. 579
Piojé	sensé			Simson, 1886: p. 267
Piratapuyo [Uaikana]	yexsépuru isébaró	yexsé yusé tsutiró		Koch-Grünberg, 1914a: p. 170, 1915–1916: p. 430
Sára—see Buhágana				
Tsöloa—see Buhágana				
Tucano	yexsépu(o)ro, yexsépuru, iéxséboro	yexsé, iéxtsé	yeyse	Koch-Grünberg, 1913: p. 960, 1915–1916: p. 430
				Pfaff, 1890: pp. 603, 605

SOUTH AMERICA (Continued)

Tribe	collared peccary	white-lipped peccary	species inadequately identified	Authority
Tuyúka	yeẓséporo	yeẓsé		Koch-Grünberg, 1914a: p. 170
Uaiana	yesépuro, yesépulo	yesé		Koch-Grünberg, 1914a: p. 170
Uaikana—see Piratapuyo				
Uanána	ye(x)sépura, yexsépuru, iaxsé	ye(x)sé, yexsextriro, iaxsé buru		Koch-Grünberg, 1914a: p. 170, 1915–1916: p. 430
Uásöna	yesémutiro	yesé		Koch-Grünberg, 1914a: p. 170
Yahúna	he̱képeraga	hehéa		Koch-Grünberg, 1914a: p. 822
Yupúa	sésebuti, (t)sé(t)sebuti	schäsché		Martius, 1863: p. 477, 1867: 2: p. 276
		sésetalage, (t)sé(t)sétalage		Koch-Grünberg, 1914a: p. 822
Ge				
Acroá [Acroamirim]		gouhobo		Martius, 1863: p. 477
Akwẽ-Shavante	uhẹ			Maybury-Lewis, 1967: p. 37
Canella [in Timbira]	klu-ré	klu		Vanzolini, 1956–1958: p. 161
Carajá [in N. Cayapó]			ichon, isá	Coudreau, 1897a: p. 264; Krause, 1911: p. 424
Cayapó (N.)	angruré	angro		Coudreau, 1897a: pp. 208, 214
	angrôre	angrô		J. B. Turner, 1967: p. 135

				Reference
Coroado	sorúcong	schórang		Martius, 1863: p. 477
Suya	angro	angro-mbedi		Seeger, 1981: p. 253
Timbira	klu-ré	klu		Lévi-Strauss, 1969: p. 86
Xikrín [in Cayapó]	angrure	angroti		Dumenil, 1975: p. 37
Panoan				
Caripuná	ono	jaua		Martius, 1867: 2: p. 241
Cashinahua	bunu yawa	yawa kuin		Kensinger, 1975: p. 27
Conebo			hondo	Farabee, 1922: p. 95
Pano			yawa	Martius, 1867: 2: p. 299
Shipibo	jono	yahua	mari, jahuá	Steinen, 1904: p. 42
				Campos, 1977: p. 59; Roe, 1982: pp. 98–99
Other languages				
Abipón			ajiranaic, cajagraye	Lafone-Quevedo, 1896: p. 216
Arikêm	tso-sesóya	tso-istá		Nimuendajú, 1932: p. 112
Auaké	dzïká(b), žïká(b)	aĩkiá, aĩ'kiá		Koch-Grünberg, 1916–1928: 4: p. 311
Aymara			kita ccuchi	Abregú Virreira, 1942: p. 254
Ayomán			monduó	Jahn, 1927: pp. 392–393
Barbacoa			melé	Lehmann, 1920: 1: p. 31
Boro [Bora]			mene	Whiffen, 1915: p. 128
Bororo	jui	jugo		Lévi-Strauss, 1969: p. 109
Botocudo	ho-kuãng	kuräck		Maximilian, 1826: pp. 558, 564
Botocudo-Crecmum	chongouïn	curähk-niptiacu niómm		Martius, 1867: 2: p. 185
Botocudo-Encreckmung	hôkuãng			Martius, 1867: 2: p. 182

Tribe	collared peccary	white-lipped peccary	species inadequately identified	Authority
		SOUTH AMERICA (Continued)		
Botocudo-Nachehe			kouraik	Martius, 1867: 2: p. 188
Botocudo-Yiporok	hok-kuéne	kourek		Martius, 1867: 2: pp. 188, 193
Carajá			ísã	Krause, 1911: p. 424
Catukina	hutschang			Martius, 1867: 2: p. 163
Camaca		kúá-hiá		Maximilian, 1826: p. 564
	kuja hatan, kuhatan	kuja, kúa-hyä		Martius, 1863: p. 477
Cayapa	aven-dú'cú	na-tcúcú, ne-tú'cú, ka'ne		Barrett, 1925: 1: p. 12
	awon-dyú'cú	tyúcú		
Chocó			pidó, pidué	Lehmann, 1920: 1: p. 94
			pidú, bidoré	
	bidó, bidoye, pidove	bidó		Mendez, 1970: pp. 239, 243
	bidobe	pido		Bennett, 1968: p. 40 (Chocó of Darién, Panama)
Coëruna		isári		Martius, 1863: p. 477, 1867: 2: p. 275
Gayon			mói	Jahn, 1927: pp. 392–393
Jijajara			monduj	Jahn, 1927: pp. 392–393
Mabenaro			vabathama	Farabee, 1922: p. 164
Makú		lewené		Koch-Grünberg, 1916–1928: 4: p. 321
Masacará		khúghah, kigha		Martius, 1863: p. 477
Mayoruna		yaúa		Martius, 1863: p. 477
Miranha	unkin		munááhä	Martius, 1867: 2: p. 278
Carapana-Tapuya				

Language				Reference
Miranha Oirá-Açu-Tapuya			*mánümö*	Martius, 1867: 2: p. 281
Mosetene	*baíi*			Armentia, 1903: p. 53
Múra		*bahũése*	*mumuni, quiti*	Nimuendajú, 1932: p. 99
Paez			*ki-ua, kutši (kutšo)*	Pittier de Fabrega, 1907: p. 348
Piaroa	*mę'kíra*		*yme* / *ímę*	Chaffanjon, 1889: p. 325 / Koch-Grünberg, 1916–1928: 4: p. 355
Puináve	*ndehüd*	*ndępí*		Koch-Grünberg, 1916–1928: 4: p. 239
Purecamecran	*croctuacuteloe*			Martius, 1863: p. 477
Sanumá [Yąnomamö]	*pose*			K. I. Taylor, 1981: p. 35
Shiriana	*bosęhʔke*	*walęke*		Koch-Grünberg, 1916–1928: 4: p. 305
Tarumá	*baiyi*	*hisu*		Farabee, 1918: p. 281
Tecuna	*hauü*	*huü*		Martius, 1863: p. 477
Toba	*mán*		*wákayi, codági*	Karsten, 1923: pp. 10, 116
Tonocoté		*pelemanpé*		Machoni de Cerdeña, 1732 [Vocab.]
Tucuna [Jucúna]	*apié*			Martius, 1867: 2: p. 253
Warrau		*ipuré*		Schomburgk, 1847–1848: 3: p. 784, 1923: 2: p. 130
Witoto			*emo, mero*	Whiffen, 1915: p. 128
Yagua	*ú·ten*	*a·wún*		Fejos, 1943: p. 122
Záparo	*cásni*	*yári*		Simson, 1886: p. 264

BIBLIOGRAPHY OF WORKS CITED

Abbeuille, Claude. 1963. *Histoire de la Mission des Pères Capvcins en l'isle de Maragnan* [1614]. Facsimile edition, Graz.

Abregú Virreira, C. 1942. *Idiomas aborígenes de la República Argentina.* Buenos Aires.

Acosta, José de. 1880. *The Natural and Moral History of the Indies* [1590]. Edited and translated by C. R. Markham. 2 vols. London: Hakluyt Society, 60, 61.

—— 1962. *Historia natural y moral de las Indias* [1590]. Edited by E. O'Gorman. México.

Acosta de Samper, S. 1894. "Los aborígenes que poblaban los territorios que hoy forman la República de Colombia en la época del descubrimiento de América." *IX Congreso Internacional de Americanistas* (Huelva, 1892), *Actas y Memorias.* 1: 373–437. Madrid.

Acuña, Cristóval de. 1698. *A Relation of the Great River of Amazons in South America* [1641] London.

—— 1859. *A New Discovery of the Great River of the Amazons* [1641] Edited and translated by C. R. Markham. London: Hakluyt Society, 24: 45–134.

Ajofrín, Francisco. 1958–1959. *Diario del viaje que por orden de la sagrada congregación de propaganda fide hizo la América septentrional en el siglo XVIII.* Edited by Vicente Castañeda y Alcover, Archivo Documental Español. 2 vols. [12, 13]. Madrid.

Albisetti, C., and A. J. Venturelli. 1962. *Enciclopedia Bororo.* 2 vols. Campo Grande.

Aldrich, J. W., and B. P. Bole. 1937. *The Birds and Mammals of the Western Slope of the Azuero Peninsula, Republic of Panama.* Cleveland: Cleveland Museum of Natural History, Scientific Publication 7.

Allen, G. M., and T. Barbour. 1923. "Mammals from Darien." *Bulletin of the Museum of Comparative Zoology* 65, 8: 259–274.

Allen, J. A. 1896. "On Mammals collected in Bexar County and vicinity, Texas, by Mr H. P. Attwater, with field notes by the collector." *Bulletin of the American Museum of Natural History* 8: 47–80.

—— 1911. "Mammals from Venezuela collected by Mr M. A. Carriker Jr." *Bulletin of the American Museum of Natural History* 30: 239–273.

Allen, P. H. 1947. "Indians of South East Colombia." *Geographical Review* 37: 567–582.

Alphonse, E. S. 1956. *Guaymi Grammar and Dictionary.* Washington, D.C.: Smithsonian Institution, Bureau of American Ethnology, Bulletin 162.

Alston, E. R. 1879–1882. *Biologia Centrali-Americana: Mammalia.* London.

Altolaguirre y Duvale, A. de. 1954. *Relaciones geográficas de la Gobernación de Venezuela, 1767–1768.* Caracas.

Alvarez de Toro, M. 1952. *Los animales silvestres de Chiapas.* Tuxtla Gutierrez: Instituto de Ciencias y Artes de Chiapas.

Alvarez López, E. 1943. "Apuntes acerca de los mamíferos americanos conocidos por Fernández de Oviedo." *Congresso de Associação Portuguesa para o Progresso das Ciências.* 5, 4 (Porto): 445–451. Porto, 1942.

Alves da Silva, Alcionilio Brüzzi. 1962. *A Civilização Indigena do Uaupes.* São Paulo.

Ambrosetti, J. B. 1894a. "Los Indios Cainguá del Alto Paraná (Misiones)." *Boletín del Instituto Geográfico Argentino* 15: 661–744.

—— 1894b. "Segundo Viage a Misiones por el Alto Paraná e Iguazú." *Boletín del Instituto Geográfico Argentino* 15: 18–114.

Ameghino, F. 1889. *Contribución al conocimiento de los mamíferos fósiles de la República Argentina.* 2 vols. Buenos Aires.

Amuchástegu, A. 1966. *Some Birds and Mammals of South America.* London.

Andagoya, Pascual de. 1829. *Relación de los sucesos de Pedrarias Dávila en las provincias de Tierra firme ó Castillo del oro, . . . escrita por el Adelantado Pascual de Andagoya* [1514–]. In: Martín Fernández de Navarrete (ed.) *Colección de viages y descubrimientos que hacieron por mar los españoles desde fin del siglo XV.* 5 vols. 1825–1837. 3: 393–456. Madrid.

—— 1865. *Narrative of the proceedings of Pedrarias Dávila in the provinces of Tierra Firme or Castilla del Oro, and the Discovery of the South Sea and the coasts of Peru and Nicaragua.* Edited and translated by C. R. Markham. London: Hakluyt Society, 34.

114

Anders, F. (intro.) 1967. *Codex Tro-Cortesianus [Codex Madrid]. True-colour facsimile-edition of the illustrated Maya-manuscript in the possession of the Museo de América, Madrid.* Codices Selecti, Graz.

Anderson, S. 1972. "Mammals of Chihuahua: taxonomy and distribution." *Bulletin of the American Museum of Natural History* 148, 2: 149–410.

Anderson, S., and J. K. Jones, eds. 1967. *Recent Mammals of the World.* New York: Ronald Press.

André, E. 1904. *A Naturalist in the Guianas.* London.

Anghiera, Pietro Martire d'. 1944. *Décadas del Nuevo Mundo* [1511–1530]. Edited and translated by Joaquín Torres Asencio. Buenos Aires.

Anon. 1858. *El Conquistador Anónimo: relación de algunas cosas de la Nueva España, y de la Gran Ciudad de Temestitán.* In: J. García Icazbalceta, ed. *Colección de Documentos para la Historia de México.* 2 vols., 1858–1866. 1: 568–598. México.

—— [? Capt. John Smith] 1882. *The Historye of the Bermudaes or Summer Islands.* Edited by J. Henry Lefroy. London: Hakluyt Society, 65.

—— 1900. *Relaciones de Yucatán.* In: J. F. Pacheco and L. Torres de Mendoza, eds. *Colección de documentos inéditos relativos al descubrimiento, conquista y organización de las antiguas posesiones españolas de ultramar.* segunda serie, 25 vols., 1885–1932. 13. Madrid.

—— 1908. *Descripción de Panamá y su Provincia* [1607]. In: M. Serrano y Sanz, ed. *Colección de libros y documentos referentes á la historia de América.* 8: 137–218. Madrid.

—— 1917. *Narrative of Some Things of New Spain and of the Great City of Temestitan, Mexico, written by the Anonymous Conqueror, a Companion of Hernán Cortés.* Edited and translated by M. H. Saville. New York: Cortes Society. 1.

—— 1963. *The Chronicle of the Anonymous Conquistador.* In: Patricia de Fuentes, editor and translator. *The Conquistadores: First-person accounts of the Conquest of Mexico,* pp. 165–181. London.

—— 1976. *Guia para visitar el zoológico de Chiapas.* Tuxtla Gutierrez: Gobierno del Estado de Chiapas.

Anthony, H. E. 1916. "Panama mammals collected 1914–1915." *Bulletin of the American Museum of Natural History* 35: 357–375.

Appun, K. F. 1961. *En los tropicos* [1849–1868]. Caracas.

Arellano Moreno, A., ed. 1964. *Relaciones geográficas de Venezuela: Recopilación, estudio preliminar y notas.* 70. Caracas: Biblioteca de la Academia Nacional de la Historia de Venezuela.

Armas, Juan Ignacio de. 1888. *La Zoología de Colón y de los primeros exploradores de América.* Habana.

Armentia, N. 1903. *Los Indios Mosetenes y su Lengua.* Buenos Aires.

Arnaud, E. 1964. "Noticia sôbre os Indios Gaviões de Oeste, Rio Tocantins, Pará." *Boletim do Museu Paraense Emílio Goeldi.* [Antropologia] 20: 1–35.

Aschmann, H. P. 1962. *Castellano-Totonaco, Totonaco-Castellano: Dialecto de la Sierra Norte de Puebla.* México.

Aschmann, H. P., and E. Dawson de Aschmann. 1973. *Diccionario Totonaco de Papantla, Vera Cruz.* México.

Asdell, S. A. 1946. *Patterns of Mammalian Reproduction.* London.

Atwood, Thomas. 1791. *The History of the Island of Dominica.* London.

Aubert de la Chesnaye-des-Bois, F. A. 1754. *Système Naturel du Règne Animal.* 2 vols. Paris.

Audubon, J. J., and J. Bachman. 1847. *The Viviparous Quadrupeds of North America.* London.

Azara, Félix de. 1801. *Essais sur l'histoire naturelle des quadrupèdes de la province du Paraguay.* 2 vols. Paris.

—— 1802. *Apuntamientos para la historia natural de los quadrupedos del Paraguay y Río de la Plata.* 2 vols. Madrid.

—— 1809. *Voyages dans l'Amérique méridionale* [1781–1801]. 4 vols. Paris.

—— 1838. *The Natural History of the Quadrupeds of Paraguay and the River La Plata.* Translated by W. P. Hunter. Edinburgh and London.

Bailey, V. 1931. *Mammals of New Mexico* [Fauna of North America 53]. Washington, D.C.: United States Department of Agriculture.

Baird, S. F. 1859. *Manual of North American Mammals, chiefly in the Museums of the Smithsonian Institution.* Philadelphia.

Baldus, H. 1931. *Indianstudien im nordöstlichen Chaco.* Leipzig.

—— 1955. "Supernatural relations with animals among Indians of eastern and southern

Brazil." *XXX International Congress of Americanists* (Cambridge, 1952). *Proceedings*. London: Royal Anthropological Institute.

—— 1970. *Tapirapé: Tribo Tupi no Brasil Central*. São Paulo.

Baltasar de Ocampo [Captain]. 1907. *Account of the Province of Vilcapampa* [1610]. Edited and translated by C. R. Markham. London: Hakluyt Society, 2d ser., 22: 203–247.

Bamburger, Joan. 1971. "The Adequacy of Kayapo Ecological Adjustment." *XXXVIII Internationalen Amerikanistenkongresses* (Stuttgart-München, 1968). *Verhandlungen* 3: 373–379. München.

Bancroft, Edward. 1769. *An Essay on the Natural History of Guiana*. London.

Bangs, O. 1901. "The Mammals collected in San Miguel Island, Panama, by W. W. Brown Jr." *American Naturalist* 35: 631–644.

—— 1902. "Chiriqui Mammalia." *Bulletin of the Museum of Comparative Zoology*. 39, 2: 17–51.

Barandiaran, D. de. 1962. "Actividades vitales de subsistencia de los indios Yekuana o Makiritare." *Antropológica* 11: 1–29.

Barbosa Rodrigues, J. 1886–1887. "Poranduba Amazonense." *Annaes da Biblioteca Nacional de Rio de Janeiro* 14, 2.

Barbot, Jean. 1732. *A Description of the Coasts of North and South Guinea* [Appendix on the Caribbee Islands]. In: A. Churchill, ed. and trans., *A Collection of Voyages and Travels*. 8 vols., 1704–1752. 5: 641–664. London.

Barker, J. 1953. "Memoria sobre la cultura de los Guaika." *Boletín Indigenista Venezolano* 1, 3–4: 433–489.

Barlow, Roger. 1932. *A Brief Summe of Geographie by Roger Barlow*. Edited by E. G. R. Taylor. London: Hakluyt Society, 2d ser., 69.

Barrère, Pierre. 1741. *Essai sur l'histoire naturelle de la France équinoxiale*. Paris.

Barrett, S. A. 1925. *The Cayapa Indians of Ecuador*. Museum of the American Indian, Indian Notes and Monographs, 40. 2 vols. New York.

Basso, Ellen B. 1973. *The Kalapalo Indians of Central Brazil*. New York.

Bayo, C. 1931. *Manual del lenguaje criollo de Centro y Sudamérica*. Madrid.

Beals, R. L. 1943. *The Aboriginal Culture of the Cáhita Indians*. Ibero-Americana, 19. Berkeley and Los Angeles.

—— 1945a. *The Contemporary Culture of the Cáhita Indians*. Washington, D.C.: Smithsonian Institution, Bureau of American Ethnology, Bulletin 142.

—— 1945b. *Ethnology of the Western Mixe*. Berkeley and Los Angeles: University of California Publications in American Archaeology and Ethnology, 42. pp. 1–175.

—— 1969. "The Tarascans." *Handbook of Middle American Indians*. Edited by E. Z. Vogt. 8: 725–773. Austin and London.

Beaumont, J. A. B. 1828. *Travels in Buenos Ayres and Adjacent Provinces of the Río de la Plata*. London.

Becerra, M. E. 1937. "Vocabulario de la Lengua Chol." *Anales del Museo Nacional* [México], 5a. época, 2: 249–278.

Beckerman, S. 1980. "Fishing and Hunting by the Barí of Colombia." In: R. B. Hames, ed. *Studies in Hunting and Fishing in the Neotropics*. Working Papers on South American Indians. Bennington College, Vermont. 2: 67–109.

Beebe, C. W. 1925. "Studies of a Tropical Jungle: One Quarter of a Square Mile of Jungle at Kartabo, British Guiana." *Zoologica* 6, 1: 5–193.

Beebe, C. W., et al. 1917. *Tropical Wild Life in British Guiana: Zoological Contributions from the Tropical Research Station of the New York Zoological Society*. New York.

Bejarana, Ignacio. 1889. *Traducción paleográfica de las actas de cabildo de la Ciudad de México*. 1 [1524–1529]. México.

Belaieff, J. 1946. "The Present-Day Indians of the Gran Chaco." In: *Handbook of South American Indians*. Edited by J. H. Steward. Washington, D.C.: Smithsonian Institution, Bureau of American Ethnology, Bulletin 143, 1: 371–380.

Bell, C. N. 1899. *Tangweera*. London.

Belt, T. 1874. *A Naturalist in Nicaragua*. London.

Benedict, F. G., and M. Steggerda. 1937. "Food of the Present-Day Maya Indians of Yucatan." Carnegie Institution, Publication 456 (Contributions to American Archaeology 3, 18), pp. 155–188.

Bennett, C. F. 1962. "The Bayano Cuna Indians, Panama: An Ecological Study of Livelihood and Diet." *Annals of the Association of American Geographers* 52: 32–50.

—— 1968a. *Human Influence on the Zoogeography of Panama.* Ibero-Americano, 51. Berkeley and Los Angeles.

—— 1968b. "Notes on Chocó Ecology in Darien Province, Panama." *Antropológica* 24: 26–55.

Benoit, P. J. 1839. *Voyage à Surinam.* Bruxelles.

Benzoni, Girolamo. 1572. *La historia del Mondo Nuovo.* Venetia.

—— 1857. *History of the New World* [1565]. Edited and translated by W. H. Smyth. London: Hakluyt Society, 21.

Bernau, J. H. 1847. *Missionary Labours in British Guiana: with Remarks on the Manners, Customs, and Superstitious Rites of the Aborigines.* London.

Bertoni, A. de W. 1915. *Fauna Paraguaya.* Asunción.

—— 1939. "Catálogos sistemáticos de los vertebradas del Paraguay." *Revista de la Sociedad Científica del Paraguay* 4, 4: 3–59.

Bewick, Thomas. 1790. *A General History of Quadrupeds.* Newcastle upon Tyne.

Bidwell, P. W., and J. I. Falconer. 1925. *History of Agriculture in the Northern United States, 1620–1860.* Washington, D.C.: Carnegie Institution.

Biet, A. 1664. *Voyage de la France équinoxiale en l'Isle de Cayenne, en l'année 1652.* Paris.

Boettger, C. R. 1958. *Die Haustiere Afrikas.* Jena.

Bolingbroke, Henry. 1807. *A Voyage to Demerary.* London and Norwich.

Boman, E. 1908. *Antiquités de la région andine de la République Argentine et du Désert d'Atacama.* 2 vols. Paris.

Borhegyi, S. F. de. 1965. "Archaeological Synthesis of the Guatemalan Highlands." *Handbook of Middle American Indians.* Edited by G. R. Willey. Austin and London. 2: 3–58.

Bourlière, F. 1955a. *The Natural History of Mammals.* Translated by H. M. Parshley. London.

—— 1955b. *Mammals of the World.* London.

—— 1973. "The Comparative Ecology of Rain Forest Mammals in Africa and Tropical America." In: B. J. Meggers, E. S. Ayensu, and D. W. Duckworth, eds. *Tropical Forest Ecosystems in Africa and South America.* Washington, D.C.: Smithsonian Institution. pp. 279–292.

Bowdich, T. E. 1821. *An Analysis of the Natural Classifications of Mammalia.* Paris.

Bowen, T. 1976. *Seri Prehistory: The Archaeology of the Central Coast of Sonora, Mexico.* Tucson: University of Arizona Anthropological Papers 27.

Boyd-Bowman, P. 1971. *Lexico Hispanoamericano del siglo XVI.* London.

Brand, D. D. 1951. *Quiroga, a Mexican Municipio.* Washington, D.C.: Smithsonian Institution, Institute of Social Anthropology, Publication 11.

—— 1964. "The Status of Ethnozoologic Studies in Meso-America." *XXXV Congreso Internacional de Americanistas* (México, 1962), Actas y Memorias 3: 131–140. México.

Brand, D. D. et al. 1960. *Coalcomán and Motines del Oro: an ex-distrito of Michaocán, México.* The Hague.

Breton, A. C. 1921. "The Aruac Indians of Venezuela." *Man* 21: 9–12.

Brett, W. H. 1851. *Indian Missions in Guiana.* London.

—— 1868. *The Indian Tribes of Guiana, their Customs and Habits.* London.

—— 1881. *Mission Work among the Indian Tribes in the Forests of Guiana.* London.

Brettes, J. de. 1903. "Les Indiens Arhouaques-Kaggabas [Arhuaco-Cagabá]." *Bulletins et mémoires de la société d'anthropologie de Paris,* sér. 5, 4: 318–357.

Brewer, F., and J. G. Brewer. 1962. *Vocabulario mexicano de Tetelcingo, Morelos.* Serie de vocabularios indígenas Mariano Silva y Aceves, 8, México.

Brigham, W. T. 1965. *Guatemala, the Land of the Quetzal.* [1st ed., 1887, New York]. Gainesville.

Brinton, D. G. 1884. "On the Language and Ethnologic Position of the Xinca Indians of Guatemala." *Proceedings of the American Philosophical Society* 22: 89–97.

Brisson, M. J. 1756. *Regnum Animale.* Paris.

Brookes, J. 1828. *A Catalogue of the Anatomical and Zoological Museum of Joshua Brookes.* London.

Browne, Patrick. 1789. *The Civil and Natural History of Jamaica.* [1st ed., 1756]. London.

Bueno, T. A. 1963. *Precolombia.* Bogotá.

Buffon, G. L. L., Comte de. 1763. *Histoire naturelle, générale et particulière.* 46 vols., 1750–1803. 10: 21–50. Paris.

—— 1812. *Natural History.* Translated by W. Smellie. 20 vols. London. 6: 404–415.

—— 1884. *Oeuvres complètes*. Edited by J. L. de Lanessan. 14 vols. Paris.

Burmeister, H. 1854–1856. *Systematische Uebersicht der Thiere Brasiliens*. 3 vols. Berlin.

—— 1876–1879. *Description physique de la République Argentine* [trad. de l'Allemand]. 5 vols. Paris.

Burney, James. 1803–1817. *A Chronological History of the Discoveries in the South Sea or Pacific Ocean*. 5 vols. London.

Burns, A. 1954. *History of the British West Indies*. London.

Burt, W. H. 1949. "Present Distribution and Affinities of Mexican Mammals." *Annals of the Association of American Geographers* 39: 211–218.

Byrd, K. M. 1976. "Changing Animal Utilization Patterns and their Implications: Southwest Ecuador 6500 B.C.–A.D. 1400." Unpublished Ph.D. thesis, University of Florida.

Cabrera, A. 1958–1961. "Catálogo de los mamíferos de América del Sur." *Revista del Instituto Nacional [Argentina] de Investigación de las Ciencias Naturales anexo al Museo: Ciencias Zoológicas* 4: 1–307, 309–732.

Cabrera, A., and J. Yepes. 1940. *Mamíferos Sud-americanos*. Tucumán and Buenos Aires.

Cadogan, L. 1966. "Animal and Plant Cults in Guaraní Lore." *Revista de Antropología* [São Paulo] 14: 105–124.

—— 1973. "Some Plants and Animals in Guaraní and Guayakí Mythology." In: J. R. Gordon (ed.) *Paraguay: Ecological Essays*. Miami. Pp. 98–104.

Cadogan, L., and M. de Colleville. 1963. "Les Indiens Guayakí de l'Yñarô [Paraguay]." *Bulletin de la Faculté des Lettres de Strasbourg* 41, 8: 439–458.

Calella, Placido de. 1942–1944. "Apuntes sobre los Indios Siona del Putumayo." *Anthropos* 35–36: 737–750.

Campos, R. 1977. "Producción de pesca y caza en una aldea Shipibo en el Río Pisqui." *Amazonia Peruana* 1, 2: 53–74.

Cappa, R. 1890. *Estudios críticos acerca de la dominación Española en América*. 20 vols. 1889–1897. Madrid. 5.

Cardim, Fernão. 1906. *A Treatise of Brazil* [1601]. In: *Purchas His Pilgrimes*. 20 vols., Glasgow. 16: 417–503.

—— 1925. *Tratados da Terra e Gente do Brasil*. Edited by B. Caetano. Rio de Janeiro.

Cardús, José. 1886. *Las misiones franciscanas entre los infieles de Bolivia*. Barcelona.

Carneiro, R. 1968. "The Transition from Hunting to Horticulture in the Amazon Basin." *VIII International Congress of Anthropological and Ethnological Sciences*. Tokyo. *Proceedings* 3: 244–248.

—— 1974. "Hunting and Hunting Magic among the Amahuaca of the Peruvian Montaña." In: P. Lyon, ed. *Native South Americans: Ethnology of the least known continent*. Boston. Pp. 122–132.

Carneiro, R., and G. E. Dole. 1956–1957. "La cultura de los Indios Kuikurus del Brasil Central." *Runa* [Buenos Aires] 8, 2: 169–202.

Carrión, Juan de [alcade mayor]. 1965. *Descripción del pueblo de Gueytlalpan* [1581]. Edited by José García Payón. Xalapa: Universidad Veracruzana.

Caspar, F. 1975. *Die Tupari. Ein Indianerstamm in Westbrasilien*. Monographien zur Völkerkunde 7. Hamburg: Museum für Völkerkunde.

Castelnau, Francis de. 1850–1857. *Expédition* [1843–1847] *dans les parties centrales de l'Amérique du Sud*. 12 vols. Paris.

Castillo, José de. 1906. *Relación de la provincia de Mojos* [1675]. Edited by M. V. Ballivián. La Paz: Documentos para la historia geográfica de la República de Bolivia. Serie primera. Época colonial, 1.

Caulín, Antonio. 1966. *Historia de la Nueva Andalucía* [1st ed., 1779]. 2 vols. Caracas.

Cervantes de Salazar, Francisco. 1971. *Crónica de la Nueva España*. Biblioteca de Autores Españoles, 2 vols. [254–255]. Madrid.

Chaffanjon, J. 1889. *L'Orenoque et le Caura*. Paris.

Chagnon, N. 1968. *Yqnomamö: The Fierce People*. New York.

Chagnon, N., and R. Hames. 1979. "Protein Deficiency and Tribal Warfare in Amazonia: New Data." *Science* 203: 910–913.

Champlain, Samuel de. 1859. *Narrative of a Voyage to the West Indies and Mexico in the years 1599 to 1602*. Edited by N. Shaw. London: Hakluyt Society, 23.

—— 1922–1936. *The Works of Samuel de Champlain*. Edited by H. P. Biggar. Toronto: Champlain Society Publications, n. s., 6 vols.

Chapman, F. M. 1929. *My Tropical Aircastle: Nature Studies in Panama*. New York.

—— 1931. "Seen from a Tropical Aircastle." *Natural History* 31, 4: 349–358.

—— 1938a. *Life in an Air Castle: Nature Studies in the Tropics.* New York and London.

—— 1938b. "White-lipped Peccary." *Natural History* 38: 408–413.

Chevalier, F. 1963. *Land and Society in Colonial Mexico: The Great Hacienda.* Berkeley and Los Angeles.

Chrostowski, N. S. 1972. "The Eco-Geographical Characteristics of the Gran Pajonal and their Relationships to some Campa Indian Cultural Patterns." *XXXIX Congreso Internacional de Americanistas* (Lima, 1970), *Actas y Memorias.* (Lima). 4: 145–160.

Cieza de León, Pedro de. 1852–1853. *La Crónica del Perú* [1532–1550]. Edited by Enrique de Vedia. Biblioteca de Autores Españoles, 2 vols. [25, 26]. Madrid.

—— 1864. *The Travels of Pedro de Cieza de León, 1532–1550, contained in the First Part of his Chronicle of Peru.* Edited and translated by C. R. Markham. London: Hakluyt Society, 33.

Cisneros, Joseph Luis de. 1764. *Descripción exacta de la provincia de Benezuela.* Valencia.

Civrieux, M. de. 1959. "Datos antropológicos de los Indios Kunuhana." *Antropológica* [Caracas] 8: 85–146.

—— 1980. "Los Cumanagoto y sus Vecinos." In: W. Coppens, ed. *Los aborigenes de Venezuela: I Etnología Antigua.* Caracas. Pp. 33–259.

Clark, J. Cooper (ed.) 1938. *Codex Mendoza and Matricula de Tributos.* 3 vols. London.

Clark, L. E., and N. Davis de Clark. 1960. *Vocabulario Popoluca-Castellano, Castellano-Popoluca: Dialecto de Sayula, Veracruz.* México: Serie de vocabularios indígenas Mariano Silva y Aceves, 4.

Clavijero, F. J. 1968. *Historia antigua de México* [ca. 1780]. Edited by R. P. M. Cuevas. México.

Cobo, Bernabé. 1956. *Obras* [*Historia del Nuevo Mundo*, 1653]. Edited by P. F. Mateos. Biblioteca de Autores Españoles, 2 vols. [91, 92]. Madrid.

Coe, M. D. 1961. *La Victoria, an early site on the Pacific coast of Guatemala.* Papers of the Peabody Museum of American Archeology and Ethnology, Harvard University, 53.

Coe, M. D., and R. A. Diehl. 1980. *In the Land of the Olmec.* 2 vols. Austin and London.

Coe, M. D., and K. V. Flannery. 1967. *Early Cultures and Human Ecology in South Coastal Guatemala.* Washington, D.C.: Smithsonian Contributions to Anthropology 3.

Collard, H., and E. Collard. 1962. *Vocabulario Mayo: Castellano-Mayo, Mayo-Castellano.* México: Serie de vocabularios indígenas Mariano Silva y Aceves, 6.

Coll y Toste, Cayetano [ed. H. W. Alberts]. 1947. "The Early History of Livestock and Pastures in Puerto Rico." *Agricultural History* 21: 61–64.

Columbus, Christopher. 1825. [*Account of fourth voyage* by Diego de Porras], and *Carta: Que escribió D. Cristóbal Colon . . . a los Christianismos y muy poderosos Rey y Reina de España.* In: Martín Fernández de Navarrete, ed. *Colección de viages y descubrimientos que hacieron por mar los españoles desde fin del siglo XV.* 5 vols. 1825–1837. Madrid. 1: 282–296, 296–313.

—— 1864. *Raccolta Completa degli Scritti di Cristoforo Colombo.* Edited by G. Battista Torre, Lione.

—— 1870. *Select Letters of Christopher Columbus, with other Original Documents relating to his Four Voyages to the New World.* Edited and translated by R. H. Major. London: Hakluyt Society, 43.

—— 1930. *The Voyages of Christopher Columbus Being the Journals of His First and Third, and the Letters concerning His First and Last Voyages to which is added the Account of His Second Voyage written by Andrés Bernaldez.* Edited and translated by C. Jane. London.

—— 1963. *Journals and Other Documents on the Life and Voyages of Christopher Columbus.* Edited and translated by S. E. Morison. New York.

—— 1969. *The Four Voyages of Christopher Columbus.* Edited and translated by J. M. Cohen. Harmondsworth.

Conzemius, E. 1927. "Die Rama-Indianer von Nicaragua." *Zeitschrift für Ethnologie* 59: 291–362.

—— 1932. *Ethnographical Survey of the Miskito and Sumu Indians of Honduras and Nicaragua.* Washington, D.C.: Smithsonian Institution, Bureau of American Ethnology, Bulletin 106.

Cooke, H. B. S., and A. F. Wilkinson. 1978. "Suidae and Tayassuidae." In: V. J. Maglio and H. B. S. Cooke, eds. *Evolution of African Mammals.* Cambridge, Mass. Pp. 435–482.

Cooper, J. M. 1949. "Traps." In: *Handbook of South American Indians.* Edited by J. H. Steward. Washington, D.C.: Smithsonian Institution, Bureau of American Ethnology, Bulletin 143, 5: 265–276.

Córdova-Rios, M., and F. B. Lamb. 1972. *The Stolen Chief.* London.

Coreal, F. 1722. *Voyages de F. Coreal aux Indes occidentales* [1666–1697]. Translated from Spanish, 3 vols. Amsterdam.

Cortés y Larraz, P. 1958. *Descripción geográfico-moral de la diócesis de Goathemala* [1768–1770]. Introduction by Adrian Recinos. 2 vols. Guatemala: Sociedad de Geografía e Historia, Guatemala.

Coudreau, H. A. 1886–1887. *La France équinoxiale: études sur les Guyanes et l'Amazonie.* 2 vols. and atlas. Paris.

—— 1897a. *Voyage au Tocantins-Araguaya.* Paris.

—— 1897b. *Voyage au Tapajoz.* Paris.

—— 1897c. *Voyage au Xingú.* Paris.

Crandall, L. S. 1964. *The Management of Wild Mammals in Captivity.* Chicago.

Crévaux, J. 1883. *Voyages dans l'Amérique du Sud.* Paris.

Crosby, A. W. 1972. *The Columbian Exchange: Biological and Cultural Consequences.* Westport.

Cuervo, A. B., ed. 1891–1894. *Colección de documentos inéditos sobre la geografía e historia de Colombia.* 4 vols. Bogotá.

Cullen, E. 1866. "The Darien Indians." *Transactions of the Ethnological Society of London,* n. s. 4: 264–267.

—— 1868. "The Darien Indians." *Transactions of the Ethnological Society of London,* n. s. 6: 150–175.

Cuvier, F. G., ed. 1804–1845. *Dictionnaire des sciences naturelles.* 60 vols. Paris.

Cuvier, G. L. C. F. D. 1817. *Le règne animal.* 4 vols. Paris.

Czekanowski, J. 1911–1924. *Forschungen im Nil–Kongo–Zwischengebiet.* 3 vols. Leipzig.

Dalquest, W. W. 1949. "The White-Lipped Peccary in the State of Veracruz, Mexico." *Anales del Instituto de Biología* [México]. 20: 411–413.

—— 1953. *Mammals of the Mexican State of San Luis Potosí.* Louisiana State University Studies, Biological Sciences Series, 1. Baton Rouge.

Dampier, William. 1906. *Dampier's Voyages.* Edited by J. Masefield. 2 vols. London.

Darlington, P. J. 1957. *Zoogeography: the Geographical Distribution of Animals.* New York.

Davis, W. B. 1944. "Notes on Mexican Mammals." *Journal of Mammalogy* 25: 370–403.

De Bry, Johann Theodor, and Johann Israel De Bry, eds. and illust. 1604. *India Orientalis 6: Regni Guineae.* Francofurti ad Moenum.

Deffontaines, P. 1957. "L'introduction du bétail en Amérique Latine." *Cahiers d'Outre-Mer* 1957: 5–22.

Delgaty, C. C. 1964. *Vocabulario Tzotzil de San Andres, Chiapas.* Serie de vocabularios indígenas Mariano Silva y Aceves, 10. México.

Denevan, W. M. 1972. "Campa Subsistence in the Gran Pajonal, Eastern Peru." *XXXIX Congreso Internacional de Americanistas.* Lima, 1970. 4: 161–179.

Dennler, J. G. 1939. "Los nombres indígenas en Guaraní de los mamíferos de la Argentina y países limítrofes y su importancia para la sistemática." *Physis* 16, 48: 225–244.

De Puydt, L. 1868. "Account of Scientific Explorations in the Isthmus of Darien in the years 1861 and 1865." *Journal of the Royal Geographical Society* 38: 69–110.

Díaz del Castillo, Bernal. 1908–1916. *The True History of the Conquest of New Spain* [ca. 1568]. Edited by Genaro García, translated by A. P. Maudslay. 2d ser., 5 vols. [23, 24, 25, 30, 40]. London: Hakluyt Society.

—— 1955. *Historia verdadera de la conquista de la Nueva España* [ca. 1568] 1944 edition. Introduction and notes by Joaquín Ramírez Cabañas. 2 vols. México.

Diego de Trujillo. 1948. *Relación del descubrimiento del reyno del Perú* [1571]. Edited by R. Porras Berrenechea. Sevilla.

Dobrizhoffer, Martin. 1822. *An Account of the Abipones, an Equestrian People of Paraguay* [1784]. 3 vols. London.

Donkin, R. A. 1977. "Spanish Red: An Ethnogeographical Study of Cochineal and the Opuntia Cactus." *Transactions of the American Philosophical Society* 67, 5.

—— n. d. *The Muscovy Duck, Cairina moschata domestica: Origins, Dispersal and Associated Aspects of the Geography of Domestication* (forthcoming).

Dorst, J. P. 1967. *South America and Central America: A Natural History.* New York.

Dreyfus, S. 1963. *Les Kayapo du nord, état de Para, Brésil: contribution à l'étude des Indiens Gé.* Paris.

Duby, G., and F. Blom. 1969. "The Lacandon." *Handbook of Middle American Indians.* Edited by E. Z. Vogt. Austin and London. 7: 276–297.

Dugès, A. 1869. "Catálogo de animales vertebrados observados en la República Mexicana." *La Naturaleza* 1: 137–145.

Dumenil, C. 1975. "Aperçu du monde animal et initiation chez les indiens Xikrin, tribu Kayapo, Brésil central." In: R. Pujol and R. Laurans, eds. *L'homme et l'animal*. Paris. Pp. 37–43.

Dumont, J. 1976. *Under the Rainbow: Nature and Supernature among the Panare Indians*. Austin.

Dureau de la Malle, M. 1855. "Cochon domestique redevenu sauvage." *Comptes rendus hebdomadaires des séances de l'Académie des Sciences* [Paris] 41: 806–807.

Dusenberry, W. H. 1963. *The Mexican Mesta: The Administration of Ranching in Colonial Mexico*. Urbana.

Dyk, Anne, and Betty Stoudt. 1965. *Vocabulario Mixteco de San Miguel el Grande*. Serie de vocabularios indígenas Mariano Silva y Aceves, 12. México.

Eckart, E. 1785. *Reise* [Brazil]. In: C. G. von Murr, ed. *Reisen einiger Missionarien der Gesellschaft Jesu in Amerika*. Nürnberg. Pp. 453–614.

Eddy, T. A. 1961. "Foods and Feeding Patterns of the Collared Peccary in Southern Arizona." *Journal of Wildlife Management* 25: 248–257.

Edmonson, M. S. 1965. *Quiche-English Dictionary*. New Orleans: Middle American Research Institute, Publication 30.

——, ed. 1971. *The Book of Counsel: The Popol Vuh of the Quiche Maya of Guatemala* [ca. 1550–1555]. New Orleans: Middle American Research Institute, Publication 35.

Edwards, Bryan. 1793. *The History, Civil and Commercial, of the British Colonies in the West Indies*. 2 vols. London.

Eisenberg, J. F., and R. W. Thorington. 1973. "A Preliminary Analysis of Neotropical Mammal Fauna." *Biotropica* 5, 3: 150–161.

Elías Ortiz, S. 1946. "Los Indios Yurumanguíes." *Acta Americana* 4: 10–25.

—— 1954. *Estudios sobre lingüística aborigen de Colombia*. Bogotá.

Elliot, D. G. 1904. *The Land and Sea Mammals of Middle America and the West Indies*. Chicago: Field Columbian Museum, Zoological Series 4, 1.

Ellisor, J. E., and W. F. Harwell. 1969. "Mobility and Home Range of the Collared Peccary in Southern Texas." *Journal of Wildlife Management* 33, 2: 425–427.

Emst, P. van. 1966. "Indians and Missionaries on the Río Tiquié, Brazil-Colombia." *Internationales Archiv für Ethnographie* 50, 2: 145–197.

Enders, R. K. 1935. "Mammalian Life Histories from Barro Colorado Island, Panama." *Bulletin of the Museum of Comparative Zoology* [Harvard]. 78: 383–502.

Enriquez de Guzman, Alonzo. 1862. *The Life and Acts of Don Alonzo Enriquez de Guzman, A.D. 1518–1543*. Edited and translated by C. R. Markham. London: Hakluyt Society, 29.

Epstein, H. 1971. *The Origin of the Domestic Animals of Africa*. 2 vols. New York and London.

Erize, E. 1960. *Diccionario comentado Mapuche-Español*. Buenos Aires.

Erxleven, J. V. P. 1777. *Systema Regni Animalis*. Lipsiae.

Espinosa, Gaspar de. 1873. *Relación del viaje que hizo desde Panamá a las provincias de París y Natá*. In: J. F. Pacheco, ed. *Colección de documentos inéditos, relativos al descubrimiento, conquista y organización de las antiguas posesiones españolas de América y Oceanía*. Madrid. 20: 5–119.

Farabee, W. C. 1917. "The Amazon Expedition of the University Museum." *Museum Journal* [University of Pennsylvania]. 8: 61–82, 126–144.

—— 1918. *The Central Arawaks*. University [of Pennsylvania] Museum, Anthropological Publications, 9. Philadelphia.

—— 1922. *Indian Tribes of Eastern Peru*. Papers of the Peabody Museum of American Archaeology and Ethnology, Harvard University, 10.

—— 1924. *The Central Caribs*. University [of Pennsylvania] Museum, Anthropological Publications, 10. Philadelphia.

Fejos, P. 1943. *Ethnography of the Yagua*. Viking Fund Publications in Anthropology, 1. New York.

Fermin, P. 1765. *Histoire naturelle de la Hollande équinoxiale*. 4 parts, continuous pagination. Amsterdam.

—— 1769. *Description générale, historique, géographique et physique de la colonie de Surinam*. 2 vols. Amsterdam.

Fernández, J. 1937–1938. *Diccionario Poconchí* [Poconchí–Spanish]. Anales de la Sociedad de Geografía e Historia de Guatemala 14: 47–70, 184–200.

Fernández de Enciso, Martín. 1948. *Suma de Geografía* [1st ed., 1519, Sevilla]. Madrid.

Fernández de Navarrete, Martín, ed. 1825–1837. *Colección de viages y descubrimientos que hacieron por mar los españoles desde fin del siglo XV.* 5 vols. Madrid.

Fernández de Oviedo y Valdés, Gonzalo. 1950. *Sumario de la natural historia de las Indias* [1526]. Edited by J. Miranda. México and Buenos Aires.

—— 1959a. *Natural History of the West Indies* [1526]. Edited and translated by S. A. Stoudemire. University of North Carolina, Studies in Romance Languages and Literatures, 32. Chapel Hill.

—— 1959b. *Historia General y Natural de las Indias* [1526–1555]. Edited by J. Pérez de Tudela Bueso, Biblioteca de Autores Españoles. 5 vols. [117–121]. Madrid.

Fernández Guardia, R. 1913. *History of the Discovery and Conquest of Costa Rica.* Translated by H. Weston Van Dyke. New York.

Fewkes, J. W. 1922. *A Prehistoric Island Culture Area of America.* Report of the Bureau of American Ethnology for 1912–1913. Washington, D.C.: Smithsonian Institution. Pp. 49–281.

Figueroa, Francisco de. 1904. *Relación de las misiones de la Compañia de Jesús en el país de los Maynas* [mid-17th century]. In: Victoriano Suárez, ed. *Colección de libros y documentos referentes á la historia de América.* 1. Madrid.

Fischer, G. 1814. *Zoognosia.* 3 vols. 1813–1814. Mosquae. 3: 284–289.

—— 1817. "Adversaria Zoologica." *Mémoires de la société impériale des naturalistes de Moscou.* 5: 357–453.

Flannery, K. V. 1967. "Vertebrate Fauna and Hunting Patterns." In: D. S. Byers, ed. *The Prehistory of the Tehuacan Valley.* Austin and London. 1: 132–177.

Flecknoe, Richard. 1654. *A Relation of Ten Years Travels.* London.

Flower, W. H., and R. Lydekker. 1891. *An Introduction to the Study of Mammals, Living and Extinct.* London.

Fock, N. 1963. *Waiwai: Religion and Society of an Amazonian Tribe.* Copenhagen: National Museum.

Forde, C. D. 1948. *Habitat, Economy and Society.* London.

Förstemann, E. 1880. *Die Maya-Handschrift der königlichen öffentlichen Bibliothek zu Dresden.* Leipzig.

—— 1902. *Commentar zur Madrider Mayahandschrift* [*Codex Tro-Cortesianus*]. Danzig.

Forsyth Major, C. I. 1897. "On the Species of *Potamochoerus,* the Bush Pigs of the Ethiopian Region." *Proceedings of the Zoological Society of London.* 1897: 359–369.

Foster, G. M. 1940. *Notes on the Popoluca of Veracruz.* México: Instituto Panamericano de Geografía e Historia, Publication 51.

—— 1942. *A Primitive Mexican Economy.* New York: American Ethnological Society.

—— 1969. "The Mixe, Zoque, Popoluca." *Handbook of Middle American Indians.* Edited by E. Z. Vogt. Austin and London. 7, 1: 448–477.

Fountain, P. 1914. *The Amazon River, from its Sources to the Sea* [1884–1885]. London.

Frantzius, A. von. 1869. "Die Säugethiere Costaricas." *Archiv für Naturgeschichte* 35, 1: 247–325.

Frenchkop. S. 1955. "Sous-ordre des Suiformes." In: P. P. Grassé, ed. *Traité de Zoologie.* Paris. 17, 1: 509–535.

Frič, V., and P. Radin. 1908. "Contribution to the Study of the Bororo Indians." *Journal of the Anthropological Institute of Great Britain and Ireland* 36: 382–406.

Friederici, G. 1947. *Amerikanistisches Wörterbuch.* Hamburg.

Frikel, P. 1968. "Os Xikrín: equipamento e técnicas de subsistência." *Museu Paraense "Emílio Goeldi," Publicações Avulsas* 7: 3–119.

Frisch, J. L. 1775. *Das Natur-System der Vierfüssigen Thiere in Tabellen.* Glogau.

Frost, M. D. 1974. "Man-Wildlife Relationship in a Tropical Lagoon Area: The case of Tortuguero (Costa Rica)." *Revista Geográfica* [México] 81: 155–164.

Fuente, Julio de la. 1947. "Los zapotecos de Choapan, Oaxaca." *Anales del Instituto Nacional de Antropología e Historia* [México] 2: 143–205.

Fuentes y Guzmán, Francisco Antonio de. 1969–1972. *Obras Históricas.* Edited by Carmelo Saenz de Santa María, Biblioteca de Autores Españoles. 3 vols. [230, 251, 259]. Madrid.

Fulop, M. 1954. "Aspectos de la cultura Tukana: cosmogonía." *Revista Colombiana de Antropología* 3: 97–137.

Gabb, W. M. 1876. "On the Indian Tribes and Languages of Costa Rica." *Proceedings of the American Philosophical Society* 14: 483–602.

—— 1881. "On the Topography and Geology of Santo Domingo." *Transactions of the American Philosophical Society* 15: 49–259.

—— 1883. "Tribus y lenguas indígenas de Costa Rica." In: D. León Fernández (ed.) *Colección de documentos para la historia de Costa Rica.* San José. 3: 303–486.

Gadow, H. 1908. *Through Southern Mexico* [1902–1904]. London.

Gann, T. W. F. 1918. *The Maya Indians of Southern Yucatan and Northern British Honduras.* Washington, D.C.

Garay, José de. 1846. *An Account of the Isthmus of Tehuantepec in the Republic of Mexico.* London.

García, Roberto Williams. 1952–1953. "Etnografía prehispanica de la zona central de Vercruz." In: I. Bernal and E. D. Hurtado, eds. *Huastecos, Totonacos y sus vecinos.* Sociedad Méxicana de Antropología. México. 13, 2 and 3: 157–161.

Garcilaso de la Vega. 1966. *Royal Commentaries of the Incas and General History of Peru* [Parts I and II]. Edited and translated by H. V. Livermore. 2 vols. Austin and London.

Gardner, G. 1846. *Travels in the Interior of Brazil* [1836–1841]. London.

Garnot, P. 1826. "Remarques sur la zoologie des îles Malouines [1822–1825]." *Annales des sciences naturelles* [Paris]. 7: 39–59.

Gates, Thomas. 1906. [*Travels of Sir Thomas Gates,* by William Strachy]. In: *Purchas His Pilgrimes.* 20 vols. Glasgow. 19: 5–72.

Gaumer, G. F. 1917. *Monografía de los mamíferos de Yucatán.* México.

Gazin, C. L. 1938. "Fossil Peccary Remains from the Upper Pliocene of Idaho." *Journal of the Washington Academy of Sciences* 28: 41–49.

Gemelli Carreri, Juan Francisco. 1704. *A Voyage round the World.* In: A. Churchill, ed. and trans. *A Collection of Voyages and Travels.* 8 vols., 1704–1752. 4: 1–606.

—— 1955. *Viaje a la Nueva España* [1700]. Edited and translated by José María de Agreda y Sánchez. 2 vols. México.

Geoffroy Saint-Hilaire, Isidore. 1861. *Acclimatation et domestication des animaux utiles.* 4th edition. Paris.

Gervais, H. P., and F. Ameghino. 1880. *Los mamíferos fosiles de la América del Sur.* Paris and Buenos Aires.

Gesner, Konrad von. 1551–1558. *Historia Animalium Libri IV.* Tiguri.

Gidley, J. W. 1920. "Pleistocene Peccaries from the Cumberland Cave Deposit." *Proceedings of the United States National Museum* 57: 651–678.

Gilbert, Humphrey. 1904. [Voyage, 1583]. In: R. Hakluyt, *The Principal Navigations, Voyages, Traffiques and Discoveries of the English Nation.* 12 vols., Glasgow. 8: 34–77.

Gill, T. 1902. "Note on the Names of the Genera of Peccaries." *Proceedings of the Biological Society of Washington* 15: 38–39.

Gillin, J. 1936. *The Barama River Caribs of British Guiana.* Papers of the Peabody Museum of American Archaeology and Ethnology, Harvard University, 14, 2.

—— 1948. "Tribes of the Guianas." In: *Handbook of South American Indians.* Edited by J. W. Steward. Washington, D.C.: Smithsonian Institution, Bureau of American Ethnology, Bulletin 143, 3: 799–860.

Gilmore, R. M. 1963 [reprint of 1st ed., 1950]. "Fauna and Ethnozoology of South America." In: *Handbook of South American Indians.* Edited by J. W. Steward. Washington, D.C.: Smithsonian Institution, Bureau of American Ethnology, Bulletin 143, 6: 345–464.

Goeje, C. H. de. 1928. *The Arawak Language of Guiana.* Amsterdam.

Goldman, E. A. 1920. *Mammals of Panama.* Washington, D.C.: Smithsonian Institution, Miscellaneous Collections, 69, 5: 1–309.

Goldman, I. 1948. "Tribes of the Uaupes-Caqueta Region." In: *Handbook of South American Indians.* Edited by J. W. Steward. Washington, D.C.: Smithsonian Institution, Bureau of American Ethnology, Bulletin 143, 3: 763–798.

—— 1963. *The Cubeo: Indians of the Northwest Amazon.* Urbana.

Goldsmith, Oliver. 1774. *An History of the Earth, and Animated Nature.* 8 vols. London.

González Dávila, Gil. 1864. *Relación de Gil González Dávila . . . de la despoblación de la Isla Española. . . .* In: S. F. Pacheco and F. Cárdenas, eds. *Colección de documentos inéditos, relativos al descubrimiento, conquista y colonización de las posesiones españolas en América y Oceanía.* 42 vols., 1864–1884. 1: 332–347. Madrid.

González Holguín, Diego. 1608. *Vocabulario de la lengua general de todo el Peru, llamada Qquichua o del Inca* [1586]. Lima.

Goodwin, G. G. 1946. "Mammals of Costa Rica." *Bulletin of the American Museum of Natural History* 87: 1–274.

Gosse, P. H. 1859. *Letters from Alabama, chiefly relating to Natural History.* London.

Gray, J. E. 1868. "Synopsis of the Species of Pigs (Suidae) in the British Museum." *Proceedings of the Zoological Society of London* 21: 17–49.

Grimes, J. E., and T. B. Hinton. 1969. "The Huichol and Cora." *Handbook of Middle American Indians.* Edited by E. Z. Vogt. Austin and London. 8: 792–813.

Gross, D. 1975. "Protein Capture and Cultural Development in the Amazon Basin." *American Anthropologist* 77: 526–549.

Groves, C. 1981. *Ancestors for the Pigs: Taxonomy and Phylogeny of the Genus Sus.* Department of Prehistory, Research School of Pacific Studies, Australian National University, Technical Bulletin 3. Canberra.

Grubb, W. B. 1911. *An Unknown People in an Unknown Land: An Account of the Lengua Indians of the Paraguayan Chaco.* London.

Guasch, A. 1961. *Diccionario Castellano–Guaraní y Guaraní–Castellano.* Sevilla.

Guenther, K. 1931. *A Naturalist in Brazil: The Fauna and Flora and the People of Brazil.* Translated by B. Miall. London.

Gumilla, José. 1791. *Historia natural, civil y geográfica de las naciones situadas en las rivieras del Río Orinoco* [1st ed., 1745, Madrid]. 2 vols. Barcelona.

Guppy, N. 1958. *Wai-Wai: Through the Forests North of the Amazon.* London.

Hackett, C. W., ed. and trans. 1923–1937. *Historical Documents relating to New Spain, Nueva Vizcaya, and Approaches thereto, to 1773* [collected by A. F. A. Bandelier and F. R. Bandelier]. 3 vols. Washington, D.C.

Haenke, T. 1901. *Descripción del Perú* [ca. 1800]. Lima.

Hahn, E. 1896. *Die Haustiere und ihre Beziehungen zur Wirtschaft des Menschen: Eine geographische Studie.* Leipzig.

Hakluyt, Richard (ed.) 1904. *The Principal Navigations, Voyages, Traffiques and Discoveries of the English Nation.* 12 vols. Glasgow.

Hall, E. R. 1946. *Mammals of Nevada.* Berkeley and Los Angeles: Museum of Vertebrate Zoology, University of California.

Hall, E. R., and W. W. Dalquest. 1963. "The Mammals of Veracruz." *Kansas University Museum of Natural History,* Publication 14: 165–362.

Hall, E. R., and K. R. Kelson. 1959. *The Mammals of North America.* 2 vols. New York.

Halley, Edmond. 1981. *The Three Voyages of Edmond Halley in the Paramore, 1698–1701.* Edited by N. J. W. Thrower, London: Hakluyt Society, 2d ser., 156.

Hamblin, Nancy Lee. 1980. "Animal Utilization by the Cozumel Maya: Interpretation through Faunal Analysis." Unpublished Ph.D. thesis, University of Arizona.

—— 1984. *Animal Use by the Cozumel Maya.* Tucson.

Hames, R. B. 1979. "A Comparison of the Efficiencies of the Shotgun and the Bow in Neotropical Forest Hunting." *Human Ecology* 7: 219–252.

—— 1980. "Game Depletion and Hunting Zone Rotation among the Ye'kwana and Yąnomamö of Amazonas, Venezuela." In: R. B. Hames (ed.) *Studies in Hunting and Fishing in the Neotropics.* Working Papers on South American Indians. Bennington College, Vermont. 2: 31–66.

Hames, R. B., and W. T. Vickers. 1982. "Optimal Diet Breadth Theory as a Model to explain Variability in Amazonian Hunting." *American Ethnologist* 9, 2: 358–378.

Hamilton, W. J. 1939. *American Mammals.* New York and London.

Harcourt, Robert. 1928. *A Relation of a Voyage* [1608–1609] *to Guiana by Robert Harcourt.* Edited by C. A. Harris. London: Hakluyt Society, 2d ser., 60.

Harner, M. J. 1972. *The Jívaro: People of the Sacred Waterfalls.* Garden City, New York.

Hartmann, R. 1884. *Die Nilländer.* Das Wissen der Gegenwart 24. Leipzig.

Hatt, R. T. 1938. "Notes concerning Animals collected in Yucatan." *Journal of Mammalogy* 19: 333–337.

Hawkes, K., K. Hill, and J. F. O'Connell. 1982. "Why Hunters Gather: Optimal Foraging and the Aché of Eastern Paraguay." *American Ethnologist* 9, 2: 379–398.

Hellmuth, N. 1977. "Cholti-Lacandon (Chiapas) and Petén-Ytzá agriculture, settlement pattern and population." In: N. Hammond, ed. *Social Process in Maya History: Studies in Honour of Sir Eric Thompson.* London. Pp. 421–448.

Helms, Mary W. 1971. *Asang: Adaptations to Culture Contact in a Miskito Community.* Gainesville.

—— 1979. *Ancient Panama: Chiefs in Search of Power.* Austin and London.

Henderson, J., and J. P. Harrington. 1914. *Ethnozoology of the Tewa Indians.* Washington, D.C.: Smithsonian Institution, Bureau of American Ethnology, Bulletin 56.

Hendey, Q. B. 1976. "Fossil Peccary from the Pliocene of South Africa." *Science* 192: 787–789.

Henley, P. 1982. *The Panare: Tradition and Change on the Amazonian Frontier.* New Haven.

Henry, J. 1964. *Jungle People: A Kaingáng Tribe of the Highlands of Brazil* [1st ed., 1941]. New York.

Hernández, Francisco. 1959. *Historia natural de Nueva España.* Translated by J. Rojo Navarro. 2 vols. [of 3 vols. of *Obras Completas*]. México.

Hernández de Alba, G. 1948a. "Sub-Andean Tribes of the Cauca Valley." In: *Handbook of South American Indians.* Edited by J. H. Steward. Washington, D.C.: Smithsonian Institution, Bureau of American Ethnology, Bulletin 143, 4: 297–327.

―― 1948b. "Tribes of the North Colombia Lowlands." In: *Handbook of South American Indians.* Edited by J. H. Steward. Washington, D.C.: Smithsonian Institution, Bureau of American Ethnology, Bulletin 143, 4: 329–338.

―― 1948c. "The Betoi and their Neighbors." In: *Handbook of South American Indians.* Edited by J. H. Steward. Washington, D.C.: Smithsonian Institution, Bureau of American Ethnology, Bulletin 143, 4: 393–398.

―― 1948d. "The Tribes of North Central Venezuela." In: *Handbook of South American Indians.* Edited by J. H. Steward. Washington, D.C.: Smithsonian Institution, Bureau of American Ethnology, Bulletin 143, 4: 475–479.

Herrera [y Tordesillas], Antonio de. 1934–1956. *Historia general de los hechos de los castellanos* [ca. 1600]. Edited by A. Ballesteros-Beretta. 15 vols. Madrid.

Hershkovitz, P. 1948. "Names of Mammals dated from Frisch, 1775, and Zimmermann, 1777." *Journal of Mammalogy* 29: 272–277.

―― 1951. "Mammals from British Honduras, Mexico, Jamaica, and Haiti." *Field Museum of Natural History, Zoology Series* (Chicago) 31: 547–569.

―― 1958. "A Geographic Classification of Neotropical Mammals." *Field Museum of Natural History, Zoology Series* (Chicago) 36: 581–620.

―― 1963. "The Nomenclature of South American Peccaries." *Proceedings of the Biological Society of Washington* 76: 85–87.

―― 1972. "The Recent Mammals of the Neotropical Region: A Zoogeographic and Ecological Review." In: A. Keast, F. C. Erk, and B. Glass, eds. *Evolution, Mammals, and Southern Continents.* Albany. Pp. 311–431.

Hill, C. A. 1966. "Peccaries, White-Lipped and Collared." *San Diego ZooNooz* 39, 7: 6–7.

Hill, John. 1752. *An History of Animals* [vol. 3 of *Natural History*, 1748–1752]. London.

Hodge, F. W., and T. H. Lewis, eds. 1907. *Spanish Explorers in the Southern United States, 1528–1543.* New York.

Hoff, B. J. 1968. *The Carib Language.* The Hague.

Hohenthal, W. 1954. "Notes on the Shucuru' Indians of Serra de Araroba, Pernambuco, Brazil." *Revista do Museu Paulista,* n. s. 8: 93–164.

Holmberg, A. 1948. "The Sirionó." In: *Handbook of South American Indians.* Edited by J. H. Steward. Washington, D.C.: Smithsonian Institution, Bureau of American Ethnology, Bulletin 143, 3: 455–463.

―― 1950. *Nomads of the Long Bow: The Sirionó of Eastern Bolivia.* Washington, D.C.: Smithsonian Institution, Institute of Social Anthropology, Publication 10.

Horton, D. 1948. "The Mundurucú." In: *Handbook of South American Indians.* Edited by J. H. Steward. Washington, D.C.: Smithsonian Institution, Bureau of American Ethnology, Bulletin 143, 3: 271–282.

Hugh-Jones, S. 1979. *The Palm and the Pleiades: Initiation and Cosmology in Northwest Amazonia.* Cambridge.

Humboldt, Alexander von. 1811. *Political Essay on the Kingdom of New Spain.* Translated by J. Black. 4 vols. London.

―― 1852–1853. *Personal Narrative of Travels to the Equinoctial Regions of America during the years 1799–1804.* Translated by T. Ross. 3 vols. London.

Hunn, E. S. 1977. *Tzeltal Folk Zoology: The Classification of Discontinuities in Nature.* San Francisco.

Hurault, J. 1968. *Les Indiens Wayana de la Guyane Française.* Paris.

Husson, A. M. 1978. *The Mammals of Suriname.* Leiden: Zoölogische Monographieën van het Rijksmuseum van Natuurlijke Historie 2.

Ibarra, J. A. 1959. *Apuntes de historia natural y mamíferos de Guatemala.* Guatemala.

Icaza, Francisco A. de. 1923. *Conquistadores y pobladores de Nueva España.* 2 vols. Madrid.

Illiger, K. 1815. "Ueberblick der Säugthiere nach ihrer Vertheilung über die Welttheile." *Abhandlungen der physikalischen Klasse der Königlich-Preusseischen Akademie der Wissenschaften* [1804–1811]. Berlin. 1815: 39–159.

Ingles, L. G. 1956. "Meat for Mayan Tables." *Pacific Discovery* 9: 4–12.

Izaguirre Ispizua, B. 1922–1929. *Historia de las misiones franciscanas y narración de los progresos de la geografía en el oriente del Perú . . . 1619–1921.* 14 vols. Lima.

Jahn, A. 1927. *Los aborígenes del occidente de Venezuela.* Caracas.

Jardine, W., ed. 1843. *The Naturalist's Library.* 40 vols., Edinburgh.

Jeffreys, Thomas. 1762. *Description of the Spanish Islands and Settlements on the Coast of the West Indies.* London.

Jena, L. S. 1944. *Popol Vuh: Das heilige Buch der Quiché-Indianer von Guatemala.* Stuttgart and Berlin.

Jewell, P. A. 1969. "Wild Animals and their Potential for New Domestication." In: P. J. Ucko and G. W. Dimbleby, eds. *The Domestication and Exploitation of Plants and Animals.* London. Pp. 101–109.

Jiménez de la Espada, M., ed. 1965. *Relaciones Geográficas de Indias.* Biblioteca de Autores Españoles, 4 vols. in 3 [183–185]. Madrid.

Johnson, A. 1977. "The Energy Costs of Technology and the Changing Environment: A Machiguenga Case." *Proceedings of the American Ethnological Society* [1975] 1977: 155–167.

Johnson, F. 1948. "The Caribbean Lowland Tribes: The Talamanca Division." In: *Handbook of South American Indians.* Edited by J. H. Steward. Washington, D.C.: Smithsonian Institution, Bureau of American Ethnology, Bulletin 143, 4: 231–251.

Johnston, H. H. 1905. "Notes on the Mammals and Birds of Liberia." *Proceedings of the Zoological Society of London* 1905 (1): 197–210.

—— 1906. *Liberia.* 2 vols. New York.

—— 1908. *George Grenfell and the Congo.* 2 vols. London and New York.

Johnstone, John. ? 1650. *Historiae Naturalis de Quadrupedibus.* Francofurti a/M.

Jones, J. K., and T. E. Lawlor. 1965. "Mammals from Isla Cozumel." *University of Kansas Museum of Natural History, Publication* 16, 3: 409–419.

Joubert, D. M., and F. N. Bonsma. 1961. "Schweinerassen in Afrika." In: J. Hammond, I. Johansson, and F. Haring, eds. *Handbuch der Tierzüchtung.* Hamburg and Berlin. 3, 2: 155–163.

Jover Peralta, A., and T. Osuna. 1952. *Diccionario Guaraní–Español y Español–Guaraní.* Buenos Aires.

Junta Colombina de México. 1893. *Vocabulario Castellano–Zapoteco* [18th century]. México.

Kahn, M. C. 1931. *Djuka: The Bush Negroes of Dutch Guiana.* New York.

Kaplan, J. 1975. *The Piaroa: A People of the Orinoco Basin.* London.

Kappler, A. 1887. *Surinam, sein Land, seine Natur, Bevölkerung und seine Kultur-Verhältnisse mit Bezug auf Kolonisation.* Stuttgart.

Karsten, R. 1920. *Contributions to the Sociology of the Indian Tribes of Ecuador.* Åbo: Acta Academiae Aboensis, Humaniora 1, 3.

—— 1923. *The Toba Indians of the Bolivian Gran Chaco.* Åbo.

—— 1926. *The Civilization of the South American Indians, with special reference to Magic and Religion.* London.

—— 1932. *Indian Tribes of the Argentine and Bolivian Gran Chaco.* Helsingfors.

—— 1935. *The Head Hunters of Western Amazonas: The Life and Culture of the Jibaro Indians of Eastern Ecuador and Peru.* Helsingfors: Commentationes Humanarum Litterarum, 7, 1.

Kaufman, T. S. 1964. "Materiales lingüísticos para el estudio de las relaciones internas y externas de la familia de idiomas Mayanos." In: E. Z. Vogt and A. Ruz, eds. *Desarrollo cultural de los Mayas.* México. Pp. 81–136.

Kelley, D. H. 1976. *Deciphering the Maya Script.* Austin and London.

Kellogg, R. 1946. "Mammals of San José Island, Bay of Panama." *Smithsonian Institution, Miscellaneous Collections* (Washington, D.C.) 106, 7: 1–4.

Kelly, I., and A. Palerm. 1952. *The Tajin Totonac: Part I—History, Subsistence, Shelter, and Technology.* Washington, D.C.: Smithsonian Institution, Institute of Social Anthropology, Publication 13.

Kensinger, K. M. 1975. "Studying the Cashinahua." In: Jane P. Dwyer, ed. *The Cashinahua of Eastern Peru.* Bristol, Rhode Island: Haffenreffer Museum of Anthropology, Pp. 9–86.

—— 1981. "Food Taboos as Markers of Age Categories in Cashinahua." In: K. M. Kensinger and W. H. Kracke, eds. *Food Taboos in Lowland South America.* Working Papers on South American Indians. Bennington College, Vermont. 3: 157–171.

Keymis, Laurence. 1904. *A Relation of the Second Voyage to Guiana Performed and Written in the Yeere 1596*. In: Richard Hakluyt, *The Principal Navigations, Voyages, Traffiques and Discoveries of the English Nation*. Glasgow. 10: 441–501.

Kidder, A. V. 1947. *The Artifacts of Uaxactun, Guatemala*. Washington, D.C.: Carnegie Institution, Publication 576.

Kidder, A. V., J. D. Jennings, and E. M. Shook. 1946. *Excavations at Kaminaljuyú, Guatemala*. Washington, D.C.: Carnegie Institution, Publication 561.

Kiltie, R. 1979. "Seed Redation and Group Size in Rain Forest Peccaries." Unpublished Ph.D. thesis, Princeton University.

—— 1980. "More on Amazon Cultural Ecology." [with reply by E. B. Ross]. *Current Anthropology* 21, 4: 541–546.

Kirchhoff, P. 1948a. "The Guayupé and Sae." In: *Handbook of South American Indians*. Edited by J. H. Steward. Washington, D.C.: Smithsonian Institution, Bureau of American Ethnology, Bulletin 143, 4: 385–391.

—— 1948b. "The Otomac." In: *Handbook of South American Indians*. Edited by J. H. Steward. Washington, D.C.: Smithsonian Institution, Bureau of American Ethnology, Bulletin 143, 4: 439–444.

—— 1948c. "Food-gathering Tribes of the Venezuelan Llanos." In: *Handbook of South American Indians*. Edited by J. H. Steward. Washington, D.C.: Smithsonian Institution, Bureau of American Ethnology, Bulletin 143, 4: 445–468.

—— 1948d. "The Tribes North of the Orinoco River." In: *Handbook of South American Indians*. Edited by J. H. Steward. Washington, D.C.: Smithsonian Institution, Bureau of American Ethnology, Bulletin 143, 4: 481–493.

—— 1948e. "The Warrau." In: *Handbook of South American Indians*. Edited by J. H. Steward. Washington, D.C.: Smithsonian Institution, Bureau of American Ethnology, Bulletin 143, 3: 869–881.

Kirkpatrick, R. D., and L. K. Sowls. 1962. "Age Determination of the Collared Peccary using the Tooth Replacement Pattern." *Journal of Wildlife Management* 26: 214–217.

Klein, Jacob Theodor. 1751. *Quadrupedum dispositio brevisque historia naturalis*. Lipsiae.

Kloos, P. 1971. *The Maroni River Caribs of Surinam*. Assen.

Koch-Grünberg, T. 1900. "Zum Animissmus der Südamerikanischen Indianer." *Internationales Archiv für Ethnographie* 13 [Supplement].

—— 1908. "Die Hianákoto-Umáua." *Anthropos* 3: 1–112.

—— 1909–1910. *Zwei Jahre unter den Indianern: Reisen in Nordwest-Brasilien, 1903–1905*. 2 vols. Berlin.

—— 1911. "Aruak-Sprachen N. W.-Brasiliens und der angrenzenden Gebiete." *Mitteilungen der Anthropologischen Gesellschaft in Wien* 41: 33–153, 203–282.

—— 1913. "Die Betóya-Sprachen Nordwest-Brasiliens und der angrenzenden Gebiete." *Anthropos* 8: 944–976.

—— 1914a. "Die Betóya-Sprachen Nordwest-Brasiliens und der angrenzenden Gebiete." *Anthropos* 9: 151–195, 569–589, 812–832.

—— 1914b. "Ein Beitrag zur Sprache der Ipuriná-Indianer." *Journal de la Société des Américanistes de Paris* n. s. 11: 57–96.

—— 1915–1916. "Die Betóya-Sprachen Nordwest-Brasiliens und der angrenzenden Gebiete." *Anthropos* 10–11: 114–158, 421–449.

—— 1916–1928. *Vom Roroima zum Orinoco: Ergebnisse einer Reise in Nord-Brasilien und Venezuela in den Jahren 1911–1913*. 5 vols. Berlin and Stuttgart.

Koch-Grünberg, T., and G. Hübner. 1908. "Die Makuschí und Wapischiána." *Zeitschrift für Ethnologie* 40: 1–44.

Kok, P. 1925–1926. "Quelques notices ethnographiques sur les indiens du Rio Papurí." *Anthropos* 20: 624–637; 21: 921–937.

Kracke, W. H. 1981. "Don't let the piranha bite your liver: A psychoanalytic approach to Kagwahiv (Tupi) food taboos." In: K. M. Kensinger and W. H. Kracke (eds.) *Food Taboos in Lowland South America*. Working Papers on South American Indians. Bennington College, Vermont. 3: 91–142.

Krause, F. 1911. *In den Wildnissen Brasiliens: Bericht und Ergebnisse der Leipziger Araaguaya-Expedition, 1908*. Leipzig.

Krieger, H. W. 1930. *The Aborigines of the Ancient Island of Hispaniola*. Report of the Bureau of American Ethnology for 1929, Washington, D.C.: Smithsonian Institution. Pp. 473–506.

Kroll, H. 1928. "Die Haustiere der Bantu." *Zeitschrift für Ethnologie* 60: 177–290.

Kruse, A. 1951–1952. *"Karusakaybë,* der Vater der Mundurukú." *Anthropos* 46: 915–932; 47: 992–1018.

Kubler, G. 1967. *The Iconography of the Art of Teotihuacan.* Studies in Pre-Columbian Art and Archaeology 4. Washington, D.C.: Dumbarton Oaks.

Labat, J. B. 1730. *Voyage du Chevalier [Renaud] des Marchais en Guinée, isles voisines, et à Cayénne, fait en 1725–1727.* 4 vols. Paris.

La Condamine, Charles Marie de. 1778. *Relation abrégée d'un voyage fait dans l'intérieur de l'Amérique méridionale.* Maestricht.

Laet, Johanne de. 1633. *Novus orbis, seu descriptionis Indiae Occidentales.* Lugduni Batavorum.

—— 1640. *L'Histoire du Nouveau Monde ou description des Indes occidentales.* Leyde.

Lafone Quevedo, S. A. 1896. *Idioma Abipón.* Buenos Aires.

Lambert, C. F. 1750. *A Collection of Curious Observations on . . . the Several Nations of Asia, Africa, and America.* Translated by J. Dunn. London.

Landa, Diego de. 1937. *Yucatan before and after the Conquest* [ca. 1566]. Edited and translated by W. Gates. Baltimore: Maya Society.

—— 1941. *Landa's relación de las cosas de Yucatán* [ca. 1566]: *a translation.* Edited and translated by A. M. Tozzer. Papers of the Peabody Museum of American Archaeology and Ethnology, Harvard University, 18.

—— 1966. *Relación de las cosas de Yucatán* [ca. 1566]. Edited by A. Garibay. México.

Landerman, P. 1973. *Vocabulario Quechua del Pastaza.* Yarinacocha, Perú.

Langdon, T. A. 1975. "Food Restrictions in the Medical System of the Barasana and Taiwano Indians of the Colombian Northwest Amazon." Unpublished Ph.D. thesis, Tulane University.

Lanning, E. P. 1967. *Peru before the Incas.* Englewood Cliffs.

Lara, J. 1971. *Diccionario Chëshwa–Castellano, Castellano–Chëshwa.* La Paz.

—— 1951. *Historia de las Indias.* Edited by A. Millares Carlo. 3 vols. México.

Las Casas, Bartolomé de. 1957. *Historia de las Indias* [1559]. Edited by J. Pérez de Tudela Bueso and E. López Oto. Madrid: Biblioteca de Autores Españoles. 2 vols. [95, 96].

—— 1958. *Apologética Historia* [1559]. Edited by J. Pérez de Tudela Bueso. Madrid: Biblioteca de Autores Españoles, 2 vols. [105, 106].

Latcham, R. E. 1922. "Los animales domesticos de la América precolombiana." *Publicaciones del Museo de Etnología y Antropología* [Santiago de Chile]. 3, 1: 1–199.

Latorre [y Setién], G., ed. 1919–1920. *Relaciones Geográficas de Indias.* Sevilla: Centro de Estudios Americanistas de Sevilla, Biblioteca Colonial Americana, 3.

—— 1920. *Relaciones Geográficas de Indias.* Sevilla: Centro de Estudios Americanistas de Sevilla, Biblioteca Colonial Americana, 4.

Laughlin, R. M. 1969. "The Huastec." *Handbook of Middle American Indians.* Edited by E. Z. Vogt. Austin and London. 7: 298–311.

Lavolat, R. 1955. "Suiformes: fossiles." In: P. Grassé, ed. *Traité de Zoologie.* Paris. 17, 1: 547–549.

Lázaro de Arregui, Domingo. 1946. *Descripción de la Nueva Galicia* [1621]. Edited by F. Chevalier. Sevilla.

Leacock, S. 1964. "Economic Life of the Maué Indians." *Boletim do Museu Paraense Emílio Goeldí* n. s. 19: 1–30.

Lefroy, J. H. 1877–1879. *Memorials of the Discovery and Early Settlement of the Bermudas or Somers Islands* [1511–1685]. 2 vols. London.

Lehmann, W. 1920. *Zentral-Amerika: Die Sprachen Zentral-Amerikas in ihren Beziehungen zueinander sowie zu Süd-Amerika und Mexiko.* 2 vols. Berlin.

Leidy, J. 1853. *The Ancient Fauna of Nebraska.* Washington, D.C.: Smithsonian Institution, Contributions to Knowledge.

Lenselink, J. 1972. "The amount of game bagged in an Amerindian Village in Surinam" [in Dutch, English summary] *De Surinaamse Landbouw* [Paramaribo]. 20, 3: 37–41.

León, N. 1912. "Vocabulario de la lengua Popoloca, Chocha o Chuchona." *Anales del Museo nacional de México,* 3a. época, 3: 1–58.

Leonardo de Argensola, Bartolomé. 1940. *Conquista de México* [ca. 1630]. From *Anales de Aragon.* Edited by J. Ramírez Cabanas. México.

León Pinelo, Antonio de. 1943. *El Paraíso en el Mundo Nuevo* [1650]. Edited by R. Porras Barrenechea. 2 vols. Lima.

Leopold, A. S. 1972. *Wildlife of Mexico: The Game Birds and Mammals* [1st ed., 1959]. Berkeley and Los Angeles.

Lerio, Joanne [Léry, Jean]. 1592. *Navigatio in Brasiliam Americae.* In: Theodor de Bry

Americae tertia pars memorabilē provinciae Brasiliae historiam continēs. Francofurti ad Moenum. Pp. 135–295.

—— 1975. *Histoire d'un voyage fait en la terre de Brésil* [1557]. Facsimile of 1580 edition. Genève.

Lerner, I. 1974. *Arcaísmos léxicos del Español de América.* Madrid.

Le Roy Gordon, B. 1957. *Human Geography and Ecology in the Sinú Country of Colombia.* Ibero-Americana, 39. Berkeley and Los Angeles.

Leunis, J. 1844–1853. *Synopsis der drei Naturreiche.* 3 vols. Hannover.

Lévi-Strauss, C. 1948. "The Tupí-Cawahíb." In: *Handbook of South American Indians.* Edited by J. H. Steward. Washington, D.C.: Smithsonian Institution, Bureau of American Ethnology, Bulletin 143, 3: 299–305.

—— 1969. *The Raw and the Cooked.* Translated by J. Weightman and D. Weightman. New York.

Lewis, R. D. 1970. "Animal Domestication in Guatemala." Unpublished M.A. thesis, University of Oregon.

Liais, E. 1872. *Climats, géologie, faune et géographie botanique du Brésil.* Paris.

Lichtenstein, H. 1830. "Erläuterungen der Nachrichten des Franc. Hernández von den vierfüssigen Thieren Neuspaniens." *Abhandlungen der Königlichen Akademie der Wissenschaften zu Berlin 1827.* 1830: 89–127.

Ligon, Richard. 1657. *A True and Exact History of the Island of Barbados.* London.

Linares, O. F. 1976a. "Garden Hunting in the American Tropics." *Human Ecology* 4, 4: 331–349.

—— 1976b. "Animals that were bad to eat were good to compete with: An analysis of the Conte style from ancient Panama." In: P. Young and J. Howe, eds. *Ritual and Symbol in Native Central America.* Eugene: University of Oregon Anthropological Papers 9. Pp. 1–20.

—— 1977. *Ecology and the Arts in Ancient Panama.* Studies in Pre-Columbian Art and Archaeology, 17. Washington, D.C.: Dumbarton Oaks Research Library and Collections.

Linares, O. F., and R. S. White. 1980. "Terrestrial Fauna from Cerro Brujo (CA-3) in Bocas del Toro and La Pitahaya (IS-3) in Chiriqui." In: O. F. Linares and A. J. Ranere, eds. *Adaptive Radiations in Prehistoric Panama.* Peabody Museum Monographs, 5, Harvard University. Pp. 181–193.

Lindskoog, J. N., and C. A. Lindskoog. 1964. *Vocabulario Cayapa.* Quito: Serie de vocabularios indígenas Mariano Silva y Aceves, 9.

Link, H. F. 1795–1797. *Beiträge zur Naturgeschichte.* 3 vols. Rostock and Leipzig.

Linnaeus, Carolus. 1758–1759. *Systema Naturae.* 10th ed., 2 vols. Holmiae.

—— 1767–1770. *Systema Naturae.* 12th ed., 3 vols. Vindobonae.

—— 1788–1793. *Systema Naturae.* 13th ed. Edited by Johann Friedrich Gmelin. 3 vols. Lipsiae.

—— 1792. *The Animal Kingdom or Zoological System of the celebrated Sir Charles Linnaeus.* Substantially a translation of Gmelin/Linnaeus by Robert Kerr. London.

Linné, J. E. S. 1929. *Darien in the Past.* Göteborg.

Lipkind, W. 1948. "The Carajá." In: *Handbook of South American Indians.* Edited by J. H. Steward. Washington, D.C.: Smithsonian Institution, Bureau of American Ethnology, Bulletin 143. 3: 179–191.

Lira, J. A. 1944. *Diccionario Kkechuwa-Español.* Tucumán.

Lizot, J. 1977. "Population, Resources, and Warfare among the Yanomami." *Man* 12, 3–4: 497–517.

—— 1979. "On Food Taboos and Amazon Cultural Ecology" [with reply by E. B. Ross]. *Current Anthropology* 20, 1: 150–155.

López de Cogolludo, Diego. 1842–1845. *Los tres siglos de la dominación española en Yucatán* [1688]. Edited by J. Sierra O'Reilly, Campeche and Mérida.

—— 1867–1868. *Historia de Yucatán* [1688]. 2 vols. Mérida.

López de Gómara, Francisco. 1954. *Historia general de las Indias con la conquista de México y de la Nueva España* [1551–1552]. Edited by P. Guibelalde and E. M. Aguilera. 2 vols. Barcelona.

—— 1964. *Cortés: The Life of the Conqueror by his Secretary* [1551–1552]. Edited and translated by L. B. Simpson. Berkeley and Los Angeles.

López de Velasco, J. 1894. *Geografía y descripción de las Indias* [1571–1574]. Edited by J. Zaragoza. Madrid.

Lothrop, S. K. 1937. *Coclé: An Archaeological Study of Central Panama.* Memoirs of the

Peabody Museum of American Archaeology and Ethnology, Harvard University, VII, 1–2.

—— 1948. "The Tribes West and South of the Panama Canal." In: *Handbook of South American Indians.* Edited by J. H. Steward. Washington, D.C.: Smithsonian Institution, Bureau of American Ethnology, Bulletin 143. 4: 253–256.

—— 1950. *Archaeology of Southern Veraguas, Panama.* Memoirs of the Peabody Museum of American Archaeology and Ethnology, Harvard University, IX, 3.

Low, D. 1842. *The Breeds of Domestic Animals of the British Islands.* London.

Lowie, R. H. 1946*a*. "The Bororo." In: *Handbook of South American Indians.* Edited by J. H. Steward. Washington, D.C.: Smithsonian Institution, Bureau of American Ethnology, Bulletin 143. 1: 419–434.

—— 1946*b*. "The Northwestern and Central Ge." In: *Handbook of South American Indians.* Edited by J. H. Steward. Washington, D.C.: Smithsonian Institution, Bureau of American Ethnology, Bulletin 143. 1: 477–517.

—— 1946*c*. "The Cariri." In: *Handbook of South American Indians.* Edited by J. H. Steward. Washington, D.C.: Smithsonian Institution, Bureau of American Ethnology, Bulletin 143. 1: 557–559.

Lozano, Pedro. 1733. *Descripción chorographica del terreno, rios, arboles y animales de las provincias del Gran Chaco.* Córdoba.

—— 1873–1874. *Historia de la conquista del Paraguay, Río de la Plata y Tucumán.* Edited by Andrés Lamas. 4 vols. Buenos Aires.

Lundelius, E. L. 1960. "*Mylohyus nasutus:* Long-nosed Peccary of the Texas Pleistocene." *Bulletin of the Texas Memorial Museum.* 1: 1–40.

Lydekker, R., and G. Blaine. 1913–1916. *Catalogue of the Ungulate Mammals in the British Museum, Natural History.* 5 vols. London.

Lyon, P. J., ed. 1974. *Native South Americans.* Boston.

Machoni de Cerdena, Antonio. 1732. *Arte y vocabulario de la lengua Tonocoté.* Madrid.

MacNeish, R. S. 1958. "Preliminary Archaeological Investigations in the Sierra de Tamaulipas, Mexico." *Transactions of the American Philosophical Society.* 48, 6.

Malbrant, R. 1952. *Faune du Centre Africain Français.* Paris.

Marcgravius, Georgius. 1648. *Historiae Rerum Naturalium Brasiliae.* Lugduni Batavorum et Amstelodami.

Marees, Pieter de. 1605. *Description et Recit Historial du Riche Royaume d'Or de Gunea.* Amsterdamme.

Mares, M. A., R. A. Ojeda, M. P. Kosko. 1981. "Observations on the Distribution and Ecology of the Mammals of Salta Province, Argentina." *Annals of the Carnegie Museum.* Washington, D.C. 50, 6: 151–206.

Markham, C. R. 1908. *Vocabularies of the General Language of the Incas of Perú or Runa Simi.* London.

Martín del Campo, R. 1941. "Ensayo de interpretación del Libro Undécimo de la Historia de Sahagún: III Los mamíferos." *Anales del Instituto de Biología* [México] 12, 1: 489–506.

—— 1960. "Contribución a la etnozoología Mixteca y Zapoteca." *Memorias y Revista de la Sociedad Científica "Antonio Alzate"* 59: 53–88.

Martínez de Yrala, Domingo. 1906. [*Letter,* 1542.] In: M. Serrano y Sanz, ed. *Colección de libros y documentos referentes á la historia de América.* Madrid. 6: 367–368.

Martius, C. F. P. von. 1863. *Glossario linguarum Brasiliensium.* Erlangen.

—— 1867. *Beiträge zur Ethnographie und Sprachenkunde Amerika's zumal Brasiliens.* 2 vols. Leipzig.

Masefield, G. B. 1967. "Crops and Livestock." In: E. E. Rich and C. H. Wilson, eds. *Cambridge Economic History of Europe.* Cambridge. 4: 275–301.

Matesanz, J. 1965. "La introducción de la ganadería en Nueva España." *Historia Mexicana* 56: 533–566.

Mather, K. F. 1922. "Exploration in the Land of the Yuracarés, Eastern Bolivia." *Geographical Review* 12: 42–56.

Maudslay, A. P. 1889–1902. *Biologia Centrali-Americana: Archaeology.* 4 vols. London.

Maximilian of Wied-Neuwied (Prince). 1820. *Travels in Brasil in the years 1815–1817.* London.

—— 1825–1832. *Beiträge zur Naturgeschichte von Brasilien.* 4 vols. Weimar.

May, Henry. 1904. *A Briefe Note of a Voyage to the East Indies* [1591–]. In: Richard Hakluyt, *The Principal Navigations, Voyages, Traffiques and Discoveries of the English Nation.* Glasgow. 10: 194–203.

Maybury-Lewis, D. 1967. *Akwẽ-Shavante Society*. Oxford.

McCullough, C. Y. 1955. "Breeding Record of Javelina, *Tayassu angulatus*, in Southern Arizona." *Journal of Mammalogy* 36: 146–149.

McDonald, D. R. 1977. "Food Taboos: A Primitive Environmental Protection Agency (South America)." *Anthropos* 72: 734–748.

McKim, F. 1947. *San Blas: An Account of the Cuna Indians of Panama* [1936]. Edited by H. Wassén. Göteborg: Etnologiska Studier, 15.

McMahon, A., and M. McMahon. 1959. *Vocabulario Cora y Español*. México: Serie de vocabularios indígenas Mariano Silva y Aceves, 2.

Méndez, E. 1970. *Los principales mamíferos silvestres de Panama*. Panama.

Menget, P. 1981. "From Forest to Mouth: Reflections on the Txicao Theory of Subsistence." In: K. M. Kensinger and W. H. Kracke, eds. *Food Taboos in Lowland South America*. Working Papers on South American Indians. Bennington College, Vermont. 3: 1–17.

Merriam, C. H. 1901a. "Six New Mammals from Cozumel Island, Yucatan." *Proceedings of the Biological Society of Washington* 14: 99–104.

—— 1901b. "Description of Four New Peccaries from Mexico." *Proceedings of the Biological Society of Washington* 14: 119–124.

Métraux, A. 1939. "Myths and Tales of the Mataco Indians." *Etnologiska Studier*. Göteborg. 9: 1–127.

—— 1942. *The Native Tribes of East Bolivia and Western Matto Grosso*. Washington, D.C.: Smithsonian Institution, Bureau of American Ethnology, Bulletin 134.

—— 1946a. "Ethnography of the Chaco." In: *Handbook of South American Indians*. Edited by J. H. Steward. Washington, D.C.: Smithsonian Institution, Bureau of American Ethnology, Bulletin 143. 1: pp. 197–370.

—— 1946b. "The Caingang." In: *Handbook of South American Indians*. Edited by J. H. Steward. Washington, D.C.: Smithsonian Institution, Bureau of American Ethnology, Bulletin 143. 1: 445–475.

—— 1948a. "Tribes of Eastern Bolivia and the Madeira Headwaters." In: *Handbook of South American Indians*. Edited by J. H. Steward. Washington, D.C.: Smithsonian Institution, Bureau of American Ethnology, Bulletin 143. 3: 381–454.

—— 1948b. "Tribes of the Eastern Slopes of the Bolivian Andes." In: *Handbook of South American Indians*. Edited by J. H. Steward. Washington, D.C.: Smithsonian Institution, Bureau of American Ethnology, Bulletin 143. 3: 465–506.

—— 1948c. "Tribes of the Jurua-Purus Basins." In: *Handbook of South American Indians*. Edited by J. H. Steward. Washington, D.C.: Smithsonian Institution, Bureau of American Ethnology, Bulletin 143. 3: 657–686.

—— 1948d. "Tribes of the Middle and Upper Amazon River." In: *Handbook of South American Indians*. Edited by J. H. Steward. Washington, D.C.: Smithsonian Institution, Bureau of American Ethnology, Bulletin 143. 3: 687–712.

—— 1948e. "The Tupinamba." In: *Handbook of South American Indians*. Edited by J. H. Steward. Washington, D.C.: Smithsonian Institution, Bureau of American Ethnology, Bulletin 143. 3: 95–133.

—— 1949a. "Weapons." In: *Handbook of South American Indians*. Edited by J. H. Steward. Washington, D.C.: Smithsonian Institution, Bureau of American Ethnology, Bulletin 143. 5: 229–263.

—— 1949b. "Boys' Initiation Rites." In: *Handbook of South American Indians*. Edited by J. H. Steward. Washington, D.C.: Smithsonian Institution, Bureau of American Ethnology, Bulletin 143. 5: 375–382.

—— 1949c. "Religion and Shamanism." In: *Handbook of South American Indians*. Edited by J. H. Steward. Washington, D.C.: Smithsonian Institution, Bureau of American Ethnology, Bulletin 143. 5: 559–599.

—— 1960. "Mythes et contes des Indiens Cayapo." *Revista do Museu Paulista* n. s. 12: 7–35.

Métraux, A., and H. Baldus. 1946. "The Guayakí." In: *Handbook of South American Indians*. Edited by J. H. Steward. Washington, D.C.: Smithsonian Institution, Bureau of American Ethnology, Bulletin 143. 1: 435–444.

Métraux, A., and C. Nimuendajú. 1946. "The Camacan Linguistic Family." In: *Handbook of South American Indians*. Edited by J. H. Steward. Washington, D.C.: Smithsonian Institution, Bureau of American Ethnology, Bulletin 143. 1: 547–552.

Middendorf, E. W. 1890. *Wörterbuch des Runa Simi oder der Keshua-Sprache*. Leipzig.

Miller, F. W. 1930. "Notes on Some Mammals of Southern Mato Grosso, Brazil." *Journal of Mammalogy* 11: 10–22.

Miller, G. S., and R. Kellogg. 1955. *List of North American Recent Mammals.* Washington, D.C.: United States National Museum, Bulletin 205.

Miller, G. S., and J. A. G. Rehn. 1902. "Systematic Results of the Study of North American Land Mammals to the close of the year 1900." *Proceedings of the Boston Society of Natural History* [1901] 30: 1–352.

Mohr, E. 1960. *Wilde Schweine.* Wittenberg Lutherstadt.

Molina, Alonso de. 1970. *Vocabulario en lengua Castellana y Mexicana* [1st ed., 1555]. México.

Molina, Giovanni Ignazio. 1809. *The Geographical, Natural, and Civil History of Chile* [1st ed., 1782]. Translated from the Italian. 2 vols. London.

Montero de Miranda, Francisco. 1953. *Descripción de la provincia de la Verapaz* [1574]. Guatemala: Anales de la Sociedad de Geografía e Historia. 27: 342–358.

Morales Padrón, Francisco. 1952. *Jamaica Española.* Sevilla.

Moran, E. F. 1979. "The Trans-Amazonica: Coping with a New Environment." In: M. L. Margolis and W. E. Carter, eds. *Brazil: Anthropological Perspectives—Essays in Honor of Charles E. Wagley.* New York. Pp. 133–159.

Moreno Toscano, A. 1968. *Geografía económica de México* [siglo XVI]. México.

Morrisey, R. J. 1957. "Colonial Agriculture in New Spain." *Agricultural History* 31, 3: 24–29.

Moseley, H. N. 1879. *Notes by a Naturalist on the "Challenger"* [1872–1876]. London.

Moser, B., and D. Taylor. 1963. "Tribes of the Piraparaná." *Geographical Journal* 129: 437–449.

Moser, E., and M. Moser. 1961. *Vocabulario Seri: Seri–Castellano, Castellano–Seri.* México: Serie de vocabularios indígenas Mariano Silva y Aceves, 5.

Muller, M.-C. 1975. "Vocabulario basico de la lengua Mapoya." *Antropológica* 42: 57–77.

Müller, P. L. S. (ed. and commentator). 1773–1776. C. Linnaeus *Vollständiges Natursystem* [*Systema Naturae*]. 8 vols. Nürnberg.

Muñoz Camargo, Diego. 1948. *Historia de Tlaxcala* [ca. 1576]. Edited by A. Chavero. México.

Muratori, L. A. 1759. *A Relation of the Missions of Paraguay.* From the French trans. of the original Italian. London.

Murie, A. 1935. *Mammals from Guatemala and British Honduras.* Ann Arbor: University of Michigan Museum of Zoology, Miscellaneous Publication 26.

Murphy, R. F. 1958. *Mundurucú Religion.* Berkeley and Los Angeles: University of California Publications in American Archaeology and Ethnology 49, 1.

—— 1960. *Headhunter's Heritage: Social and Economic Change among the Mundurucú Indians.* Berkeley and Los Angeles.

Murphy, R. F., and Y. Murphy. 1974. *Women of the Forest.* New York.

Murphy, R. F., and B. Quain. 1955. *The Trumai Indians of Central Brazil.* New York: American Ethnological Society, Monograph 24.

Murra, J. 1948. "The Cayapa and Colorado." In: *Handbook of South American Indians.* Edited by J. H. Steward. Washington, D.C.: Smithsonian Institution, Bureau of American Ethnology, Bulletin 143. 4: 277–291.

Myers, T. 1976. "Defended Territories and No-Man's Lands." *American Anthropologist* 78: 354–355.

Nader, Laura. 1969a. "The Zapotec of Oaxaca." *Handbook of Middle American Indians.* Edited by E. Z. Vogt. Austin and London. 7, 1: 329–359.

—— 1969b. "The Trique of Oaxaca." *Handbook of Middle American Indians.* Edited by E. Z. Vogt. Austin and London. 7, 1: 400–416.

Nathusius, H. von. 1864. *Vorstudien für Geschichte und Zucht der Hausthiere Zunæchst am Schweinschædel.* Berlin.

Navarrete, Domingo. 1962. *The Travels and Controversies of Friar Domingo Navarrete, 1618–1686.* Edited by J. S. Cummins. London: Hakluyt Society. 2nd ser., 2 vols. [118, 119].

Neal, B. J. 1959. "A Contribution on the Life History of the Collared Peccary in Arizona." *American Midland Naturalist* 61: 177–190.

Neel, J. V., F. M. Salzano, P. C. Junqueira, F. Keiter, and D. Maybury-Lewis. 1964. "Studies on the Xavante Indians of the Brazilian Mato Grosso." *American Journal of Human Genetics* 16, 1: 52–140.

Nelson, E. A. 1916. "The Larger North American Mammals." *National Geographic Magazine* 30, 5: 385–472.

Neumann, P. 1967. *Wirtschaft und materielle Kultur der Buschneger Surinames.* Dresden: Abhandlungen und Berichte des Staatlichen Museums für Völkerkunde Dresden, 26.

Nietschmann, B. 1972*a*. "Hunting and Fishing Focus among the Miskito Indians, Eastern Nicaragua." *Human Ecology* 1, 1: 41–67.

—— 1972*b*. "Hunting and Fishing Productivity of the Miskito Indians, Eastern Nicaragua." *XXXIX Congreso Internacional de Americanistas* (Lima, 1970), *Actas y Memorias* (Lima) 4: 69–88.

—— 1973. *Between Land and Water: The Subsistence Ecology of the Miskito Indians, Eastern Nicaragua.* New York and London.

Nimis, Marion M. 1982. "The Contemporary Role of Women in Lowland Maya Livestock Production." In: K. V. Flannery (ed.) *Maya Subsistence.* New York and London. Pp. 313–325.

Nimuendajú, C. 1915. "Sagen der Tempé-Indianer." *Zeitschrift für Ethnologie* 47: 281–305.

—— 1919–1922. "Bruchstücke aus Religion und Überlieferung der Šipáia-Indianer." *Anthropos* 14–15 [1919–1920]: 1002–1038; 16–17 [1921–1922]: 367–406.

—— 1932. "Wortlisten aus Amazonien." *Journal de la Société des Americanistes de Paris* n. s. 24: 93–119.

—— 1939. *The Apinayé.* Edited and translated by R. H. Lowie and J. M. Cooper. Washington, D.C.

—— 1946. *The Eastern Timbira.* Berkeley and Los Angeles: University of California Publications in American Archaeology and Ethnology, 41.

—— 1948*a*. "Tribes of the Lower and Middle Xingú River." In: *Handbook of South American Indians.* Edited by J. H. Steward. Washington, D.C.: Smithsonian Institution, Bureau of American Ethnology, Bulletin 143. 3: 213–243.

—— 1948*b*. "The Cawahib, Parintintin, and their Neighbors." In: *Handbook of South American Indians.* Edited by J. H. Steward. Washington, D.C.: Smithsonian Institution, Bureau of American Ethnology, Bulletin 143. 3: 283–297.

—— 1948*c*. "The Mura and Piraha." In: *Handbook of South American Indians.* Edited by J. H. Steward. Washington, D.C.: Smithsonian Institution, Bureau of American Ethnology, Bulletin 143. 3: 255–269.

—— 1948*d*. "The Cayabi, Tapanyuna, and Apiacá." In: *Handbook of South American Indians.* Edited by J. H. Steward. Washington, D.C.: Smithsonian Institution, Bureau of American Ethnology, Bulletin 143. 3: 307–320.

—— 1952. *The Tukuna.* Edited by R. H. Lowie, translated by W. Hohenthal. Berkeley and Los Angeles.

Nino, Bernardino de. 1913. *Guia al Chaco Boliviano.* La Paz.

Nordenskiöld, E. 1912. *La vie des Indiens dans le Chaco.* Translated by H. Beuchat. Paris.

—— 1915. *Forskningar och äventyr i Sydamerika.* Stockholm.

—— 1920. *The Changes in the Material Culture of Two Indian Tribes under the Influence of New Surroundings.* Göteborg: Comparative Ethnographic Studies, 2.

—— 1924. *Forschungen und Abenteuer in Südamerika.* Stuttgart.

Núñez Cabeza de Vaca, Alvar. 1891. *The Commentaries* [1541–1544] *of Alvar Núñez Cabeza de Vaca by Pero Hernández* [Secretary to the Adelantado]. Edited and translated by L. L. Dominguez. London: Hakluyt Society, 81.

—— 1906. *Relación de los naufragios y comentarios de Alvar Núñez Cabeza de Vaca.* 2 vols. Madrid.

Nuttall, Z. (intro.) 1902. *Codex Nuttall: Facsimile of an Ancient Mexican Codex.* Peabody Museum of American Archaeology and Ethnology, Harvard University, Cambridge.

Oberg, K. 1949. *The Terena and the Caduveo of the Southern Mato Grosso, Brazil.* Washington, D.C.: Smithsonian Institution, Institute of Social Anthropology, Publication 9.

—— 1953. *Indian Tribes of Northern Mato Grosso, Brazil.* Washington, D.C.: Smithsonian Institution, Institute of Social Anthropology, Publication 15.

Olrog, C. C., R. A. Ojeda, and R. M. Barquez. 1976. "*Catagonus wagneri* (Rusconi) en el noroeste Argentino." *Neotrópica* 22: 53–56.

Olsen, S. J. 1964. *Mammal Remains from Archaeological Sites: Part I, Southeastern and Southwestern United States.* Papers of the Peabody Museum of American Archaeology and Ethnology, Harvard University, 61, 2.

—— 1972. "Animal Remains from Altar de Sacrificios." In: G. R. Willey, ed. *Artifacts of*

Altar de Sacrificios. Papers of the Peabody Museum of American Archaeology and Ethnology, Harvard University, 64, 1: 243–246.

—— 1978. "Vertebrate Faunal Remains." In: G. R. Willey, ed. *Excavations at Seibal.* Memoirs of the Peabody Museum of American Archaeology and Ethnology, Harvard University, 14, 1: 172–176.

—— 1982. *An Osteology of Some Maya Mammals.* Papers of the Peabody Museum of American Archaeology and Ethnology, Harvard University, 73.

Orbigny, Alcide Dessalines d'. 1835–1847. *Voyage dans l'Amérique méridionale* [1826–1833]. 9 vols. Paris.

—— 1845. *Descripción geográfica, histórica y estadística de Bolivia.* Paris.

Orr, C., and B. Wrisley. 1965. *Vocabulario Quichua del Oriente de Ecuador.* Quito: Serie de vocabularios indígenas Mariano Silva y Aceves, 11.

Ortiz Mayans, A. 1932. *Diccionario Guaraní-Castellano, Castellano-Guaraní.* Asunción.

Osgood, W. H. 1914a. *Mammals of an Expedition across Northern Peru.* Chicago: Field Museum of Natural History, Zoological Series, 10: 143–185.

—— 1914b. *Mammals from Western Venezuela and Eastern Colombia.* Chicago: Field Museum of Natural History, Zoological Series, 10: 33–66.

—— 1921. "Notes on Nomenclature of South American Mammals." *Journal of Mammalogy* 2: 39–40.

Paez, R. 1868. *Travels and Adventures in South and Central America.* London.

Palacios, E. J. 1928. *En los confines de la selva Lacandóna: Exploraciones en el estado de Chiapas, Mayo-Agosto, 1926.* México.

Palmer, T. S. 1897. "Notes on the Nomenclature of Four Genera of Tropical American Mammals." *Proceedings of the Biological Society of Washington* 11: 173–174.

Park, W. Z. 1947. "Tribes of the Sierra Nevada de Santa Marta, Colombia." In: *Handbook of South American Indians.* Edited by J. H. Steward. Washington, D.C.: Smithsonian Institution, Bureau of American Ethnology, Bulletin 143. 2: 865–886.

Parsons, J. J. 1962. "The Acorn-Hog Economy of the Oak Woodlands of South West Spain." *Geographical Review* 52: 211–235.

Paso y Troncoso, Francisco (ed.) 1905a. *Papeles de Nueva España: Geografía y Estadística,* 4 [*Relaciones Geográficas de la Diócesis de Oaxaca, 1579–1581*]. Madrid.

—— 1905b. *Papeles de Nueva España: Geografía y Estadística,* 5 [*Relaciones Geográficas de la Diócesis de Tlaxcala, 1580–1582*]. Madrid.

—— 1979. *Relaciones geográficas de México* [*Papeles de Nueva España: Geografía y Estadística—Relaciones Geográficas de la Diócesis de México, 1579–1582, México, 1890*]. México.

Paz Soldan, D. D. M. 1862. *Geografía del Perú.* Paris.

Pennington, C. W. 1969. *The Tepehuan of Chihuahua: Their Material Culture.* Salt Lake City.

—— 1979–1980. *The Pima Bajo of Central Sonora, Mexico.* 2 vols. [1, *Material Culture,* 1980, 2, *Vocabulario en la lengua Nevome,* 1979]. Salt Lake City.

Pereyra, C. 1920. *La obra de España en América.* Madrid.

Perry, R. 1970. *The World of the Jaguar.* Newton Abbot.

Pfaff, F. 1890. "Die Tucanos des oberen Amazonas." *Zeitschrift für Ethnologie* 22: 597–606.

Pfefferkorn, Ignaz. 1949. *Sonora: A Description of the Province* [1794–1795]. Translated by T. E. Treutlein, 2 parts in 1 vol. Albuquerque.

Phillips, J. F. V. 1926. "Wild Pig (*Potamochoerus choeropotamus*) at the Knyasna." *South African Journal of Science* 23: 655–660.

Pickett, Velma B. 1959. *Vocabulario Zapoteco del Istmo.* México: Serie de vocabularios indígenas Mariano Silva y Aceves, 3.

Pierret, P. V., and M. J. Dourojeanni. 1966. "La caza y la alimentación humana en las riberas del Río Pachitea, Perú." *Turrialba* 16: 271–277.

—— 1967. "Importancia de la caza para alimentación humana en el curso inferior del Río Ucayali, Perú." *Revista Forestal del Perú* 1: 10–21.

Pina Chan, R. 1971. "Preclassic or Formative Pottery and Minor Arts of the Valley of Mexico." *Handbook of Middle American Indians.* Edited by G. F. Ekholm and I. Bernal. Austin and London. 10: 157–178.

Pineda, E. 1845. *Descripción geográfica del departamento de Chiapas y Soconusco.* México.

Piso, Gulielmus. 1658. *Historiae naturalis et medicae Indiae Occidentalis.* Amstelaedami.

Pittier de Fábrega, H. 1907. "Ethnographic and Linguistic Notes on the Paez Indians of Tierra Adentro, Cauca, Colombia." *Memoirs of the American Anthropological Association* 1: 301–356.

Plagemann, A. 1888. "Ausflüge in die Cordilleren der Hacienda de Cauquenes." *Verhandlungen des Deutschen wissenschaftlichen Vereins zu Santiago de Chile* 6: 277–323.

Pohl, Mary E. D., and L. H. Feldman. 1982. "The Traditional Role of Women and Animals in Lowland Maya Economy." In: K. V. Flannery, ed. *Maya Subsistence*. New York and London. Pp. 295–311.

Pollock, H. E. D., and C. E. Ray. 1957. "Notes on the Vertebrate Animal Remains from Mayapan." Washington, D.C.: Carnegie Institution, Current Reports, 41: 633–656.

Pollock, H. E. D., R. L. Roys, and T. Proskouriakov. 1962. *Mayapan, Yucatan, Mexico.* Washington, D.C.: Carnegie Institution, Publication 619.

Pomar, Juan Bautista. 1891. *Relación de Texcoco* [1582]. In: J. G. Icazbalceta, ed. *Nueva colección de documentos para la historia de México.* México. 3: 1–69.

Ponce, Alonso. 1875. *Relación breve y verdadera de algunas cosas de las muchas que sucedieron al padre fray Alonso Ponce en las provincias de la Nueva España* [ca. 1590]. In: M. Salvá and Marques de la Fuensanta de Valle, eds. *Colección de documentos inéditos para la historia de España,* 2 vols. [57, 58]. Madrid.

—— 1932. *Fray Alonso Ponce in Yucatán, 1588.* Edited and translated by E. Noyes. New Orleans: Middle American Research Institute, Publication 4.

Prado, C. 1967. *The Colonial Background of Modern Brazil.* Berkeley and Los Angeles.

Puente y Olea, Manuel de la. 1900. *Los trabajos geográficos de la Casa de Contratación.* Sevilla.

Quandt, C. 1807. *Nachricht von Suriname.* Görlitz.

Ralegh, Walter. 1848. *The Discovery of the Large, Rich and Beutiful Empire of Guiana* [1595]. Edited by R. H. Schomburgk. London: Hakluyt Society, 3.

Ranere, A. J. 1980. "Preceramic Shelters in the Talamancan Range." In: O. F. Linares and A. J. Ranere, eds. *Adaptive Radiations in Prehistoric Panama.* Peabody Museum of American Archaeology and Ethnology, Monograph 5 (Harvard University), pp. 16–43.

Ray, John. 1693. *Synopsis Methodica Animalium Quadrupedum.* Londini.

Real Academia Española. 1970. *Diccionario de la lengua Española.* Madrid.

Reclus, E. 1881. *Voyage à la Sierra Nevada de Sainte-Marthe.* Paris.

Redfield, R. 1942. *The Folk Culture of Yucatan.* 2nd imp., Chicago.

Redfield, R., and A. Villa Rojas. 1962. *Chan Kom: A Maya Village.* 1st edition 1934, Carnegie Institution, Washington, D.C. Chicago.

Regehr, W. 1979. *Die lebensräumliche Situation der Indianer im paraguayischen Chaco.* Basel: Basler Beiträge zur Geographie.

Reichel-Dolmatoff, G. 1971. *Amazonian Cosmos.* Chicago.

—— 1975. *The Shaman and the Jaguar.* Philadelphia.

—— 1976. "Cosmology as Ecological Analysis: A View from the Rain Forest." *Man* 11, 3: 307–318.

Reichenbach, A. B. 1835. *Bildergallerie der Thierwelt.* Leipzig.

Reinhardt, J. 1869. [Letter from Prof. J. Reinhardt to the Secretary of the Zoological Society of London, Copenhagen, 1869] *Proceedings of the Zoological Society of London* 1869: 56–57.

Rengger, J. R. 1830. *Naturgeschichte der Säugetiere von Paraguay.* Basel.

—— 1835. *Reise nach Paraguay in den Jahren 1818–1826.* Aarau.

Restrepo Tirado, E. 1892a. *Ensayo etnográfico y arqueológico de la provincia de los Quimbayas en el Nuevo Reino de Granada.* Bogotá.

—— 1892b. *Estudios sobre los aborígenes de Colombia.* Bogotá.

Reynoso, Diego de. 1897. "Vocabulario en lengua Mame [1644]." *Actes de la Société Philologique* [Paris] 25: 267–351.

Ribeiro, D. 1950. *Religião e mitologia Kadiuéu.* Rio de Janeiro: Serviço do Proteção aos Índios, Publicação 106.

—— 1951. "Noticia dos Ofaié-Chavante." *Revista do Museu Paulista* n. s. 5: 105–135.

Ricardo, Antonio. 1970. *Gramatica Quechua y Vocabularios* [adaptación de la primera edición de la obra de Antonio Ricardo, 1586, by Rafael Aguilar Paez]. Lima.

Rice, H. 1910. "The River Uaupés." *Geographical Journal* 35: 682–700.

Ricketson, O. G. and E. B. Ricketson. 1937. *Uaxactun, Guatemala: Group E, 1926–1931.* Washington, D.C.: Carnegie Institution, Publication 477.

Roberts, George (pseudonym). 1726. *The Four Years Voyage of George Roberts.* London.

Roberts, O. 1827. *Narrative of Voyages and Explorations on the East Coast and in the Interior of Central America.* Edinburgh.

Robertson, J. A. 1927. "Some Notes on the Transfer of Plants and Animals by Spain to its

Colonies Overseas." In: W. W. Pierson, ed. *Studies in Hispanic-American History.* James Sprunt Historical Studies. 19, 2. Chapel Hill: University of North Carolina.

Rochefort, Charles de. 1666. *Relation de l'isle de Tabago.* Paris.

Rochefort, C. de, and L. de Poincy. 1658. *Histoire naturelle et morale des iles Antilles de l'Amérique.*

—— 1666. *The History of the Caribby Islands.* Translated by J. Davies. London.

Rodríguez Herrera, E. 1959. *Léxico mayor de Cuba.* 2 vols. La Habana.

Roe, P. G. 1982. *The Cosmic Zygote: Cosmology in the Amazon Basin.* New Brunswick.

Röhl, E. 1959. *Fauna descriptiva de Venezuela: Vertebrades.* Madrid.

Rohlfs, G. F. 1874–1875. *Quer durch Afrika.* 2 vols. Leipzig.

Rojas González, Francisco. 1949. "Los Zapotecos en la época prehispánica." In: L. Mendieta y Núñez, ed. *Los Zapotecos.* México. Pp. 35–104.

Romero, E. 1959. *Historia económica del Perú.* Buenos Aires.

Roots, C. G. 1966. "Notes on the Breeding of White-Lipped Peccaries at Dudley Zoo." *International Zoological Yearbook* 6: 198–199.

Ross, E. B. 1978. "Food Taboos, Diet, and Hunting Strategy: The Adaptation to Animals in Amazon Cultural Ecology." *Current Anthropology* 19, 1: 1–36.

Roth, W. E. 1915. *An Inquiry into the Animism and Folklore of the Guiana Indians.* Report of the Bureau of American Ethnology for 1908–1909, pp. 103–386. Washington, D.C.: Smithsonian Institution.

—— 1924. *An Introductory Study of the Arts, Crafts and Customs of the Guiana Indians.* Report of the Bureau of American Ethnology for 1916–1917, pp. 27–745. Washington, D.C.: Smithsonian Institution.

Roulin, M. 1835. "Recherches sur quelques changemens observés dans les animaux domestiques transportés de l'ancien dans le nouveau continent." *Mémoires présentés par divers savans à l'Académie Royale des Sciences de l'Institut de France* 6: 321–352.

Roys, R. L. 1933. *The Book of Chilam Balam of Chumayel.* Edited and translated by R. L. Roys. Washington, D.C.: Carnegie Institution, Publication 438.

—— 1943. *The Indian Background of Colonial Yucatan.* Washington, D.C.: Carnegie Institution, Publication 548.

Ruddle, K. 1970. "The Hunting Technology of the Maracá Indians." *Antropológica* [Caracas] 25: 21–63.

Rugendas, Johann Moritz. 1835. *Voyage pittoresque dans le Brésil.* Paris.

Ruiz Blanco, Matías. 1888. *Arte y tesora de la lengua Cumanagota.* In: J. Platzmann, ed. *Algunas obras raras sobre la lengua Cumanagota.* Leipzig. 3.

—— 1965. *Conversión de Píritu* [ca. 1690]. Edited by F. de Lejarza, Biblioteca de la Academia Nacional de la Historia [Fuentes para la Historia Colonial de Venezuela] 78, Caracas.

Ruiz de Montoya, Antonio. 1876. *Gramatica y diccionarios* [arte, vocabulario, tesoro] *de la lengua Tupi ó Guarani* [1639]. Paris and Vienna.

Rusconi, C. 1931a. "Las especies fósiles argentinas de pecaríes ("Tayassuidae") y sus relaciones con las del Brasil y Norte América." *Anales del Museo Nacional de Historia Natural "Bernardino Rivadavia"* 36: 121–227.

—— 1931b. "La presencia del género "Platygonus" en túmulos indígenas de época prehispánica." *Anales del Museo Nacional de Historia Natural "Bernardino Rivadavia"* 36: 228–241.

Rye, W. B. (ed.) 1851. *The Discovery and Conquest of Terra Florida by Don Fernando de Soto* [ca. 1540]. Written by a gentleman of Elvas, translated from the Portuguese by Richard Hakluyt. London: Hakluyt Society, 9.

Sack [Zak], Albert von. 1810. *A Narrative of a Voyage to Surinam* [1805–1807]. Translated from the German. London.

Sahagún, Bernardino de. 1963. *Florentine Codex: General History of the Things of New Spain* [ca. 1570]. Edited and translated by C. E. Dibble and A. J. O. Anderson. Book 11. Santa Fé.

Salle, A. 1857. "Liste des oiseaux rapportés et observés dans la République Dominicaine [1849–1851]." *Proceedings of the Zoological Society of London* 1857: 230–237.

Sanborn, C. C. 1949. "Mammals from the Río Ucayali, Peru." *Journal of Mammalogy* 30: 277–288.

Sánchez Labrador, José. 1910. *El Paraguay católico* [ca. 1766]. 2 vols. Buenos Aires.

Sanderson, I. T. 1950. "A Brief Review of the Mammals of Suriname (Dutch Guiana) based upon a collection made in 1938." *Proceedings of the Zoological Society of London* 119, 2: 755–789.

Santamaría, F. J. 1942. *Diccionario general de americanismos.* 3 vols. México.
—— 1959. *Diccionario de mejicanismos.* Méjico.
Santiago Bertoni, M. 1973. "The Conditions of Animal Life in Paraguay." In: J. R. Gorham, ed. *Paraguay: Ecological Essays,* pp. 65–69. Miami.
Saravia, Albertina. 1969. *Popol Vuh: Illustradas con dibujos de los codices Mayas.* México.
Sargot, P. 1882. "Vocabulaire François–Arrouague." In: J. Crevaux, P. Sargot, and L. Adam. *Grammaires et vocabulaires Roucouyenne, Arrouague, Piapoco, et d'autres langues de la région des Guyanes,* pp. 61–68. Paris.
Sauer, C. O. 1966. *The Early Spanish Main.* Berkeley and Los Angeles.
Schmidt, M. 1905. *Indianerstudien in Zentralbrasilien.* Berlin.
Schmidt, Ulrich. 1891. *Voyage of Ulrich Schmidt to the Rivers La Plata and Paragual* [1535–1552]. Edited and translated by L. L. Dominguez. London: Hakluyt Society, 81.
Schoenhals, A., and L. C. Schoenhals. 1965. *Vocabulario Mixe de Totontepec.* México: Serie de vocabularios indígenas Mariano Silva y Aceves, 14.
Scholes, F. V., and R. L. Roys. 1948. *The Maya Chontal Indians of Acalan-Tixchel: A Contribution to the History and Ethnography of the Yucatan Peninsula.* Washington, D.C.: Carnegie Institution, Publication 560.
Schomburgk, R. H. 1836. "Report of an Expedition into the Interior of British Guayana in 1835–1836." *Journal of the Royal Geographical Society* 6: 224–284.
—— 1837. "Diary of an Account of the River Berbice in British Guayana in 1836–1837." *Journal of the Royal Geographical Society* 7: 302–350.
—— 1841. "Journey from Fort San Joaquin, on the Río Branco, to Roraima, and thence by the Rivers Parima and Merewari to Esmeralda, on the Orinoco, in 1838–1839." *Journal of the Royal Geographical Society* 10: 191–247.
—— 1847–1848. *Reisen in Britisch-Guiana in den Jahren 1840–1844.* 3 vols. Leipzig.
—— 1923. *Travels in British Guiana* [1840–1844]. Edited and translated by W. E. Roth. 2 vols. Georgetown.
Schreber, J. C. D. von. 1778–1855. *Die Säugthiere in Abbildungen nach der Natur mit Beschreibungen.* 7 vols. and 3 vols. of Supplement by J. A. Wagner. Erlangen.
Schwab, G. 1947. *Tribes of the Liberian Hinterland.* Papers of the Peabody Museum of American Archaeology and Ethnology, Harvard University, 31.
Schweinfurth, G. 1873. *The Heart of Africa.* 2 vols. London.
—— 1918. *Im Herzen von Afrika* [1868–1871]. Leipzig.
Schweinsberg, R. E. 1971. "Home Range, Movements and Herd Integrity of the Collared Peccary." *Journal of Wildlife Management* 35: 455–460.
Schweinsberg, R. E., and L. K. Sowls. 1972. "Aggressive Behaviour and Related Phenomena in the Collared Peccary." *Zeitschrift für Tierpsychologie* 30: 132–143.
Sclater, P. L. 1860. [*Report*] *Proceedings of the Zoological Society of London* 1860: 443.
Seeger, A. 1981. *Nature and Society in Central Brazil: The Suya Indians of Mato Grosso.* Cambridge, Mass., and London.
Seler, E. 1902–1923. *Gesammelte Abhandlungen zur Amerikanischen Sprach und Altertumskunde.* vol. 5. Berlin.
—— 1909. "Die Tierbilder der mexikanischen und Maya-Handschriften." *Zeitschrift für Ethnologie* 41: 209–257, 381–457, 784–846.
Sick, H. 1959. *Tukani.* Translated by R. H. Stevens. London.
Sillar, F. C., and R. M. Meyer. 1961. *The Symbolic Pig: An Anthology of Pigs in Literature and Art.* Edinburgh and London.
Silverman-Cope, P. 1973. "A Contribution to the Ethnography of the Colombian Makú." Unpublished Ph.D. thesis, Cambridge University.
Simón, Pedro. 1627. *Noticias historiales de las conquistas de Tierra Firme: Primera Parte.* Madrid.
—— 1882–1892. *Noticias historiales de las conquistas de Tierra Firme.* 5 vols. Bogotá.
Simons, F. A. A. 1885. "An Exploration of the Goajira Peninsula, United States of Colombia." *Proceedings of the Royal Geographical Society* n. s. 7: 781–796.
Simoons, F. J. 1953. "Notes on the Bush Pig (*Potamochoerus*)." *Uganda Journal* 17: 80–81.
—— 1961. *Eat Not This Flesh: Food Avoidances in the Old World.* Madison.
Simpson, G. G. 1941. "Vernacular Names of South American Mammals." *Journal of Mammalogy* 22: 1–17.
Simson, A. 1886. *Travels in the Wilds of Ecuador and Explorations of the Putumayo River.* London.
Siskind, Janet. 1973. *To Hunt in the Morning.* New York.

Skutch, A. F. 1971. *A Naturalist in Costa Rica.* Gainesville.

Sloane, Hans. 1707–1725. *A Voyage to the Islands of Madera, Barbados, Nieves, S. Christopher's, and Jamaica, with the Natural History of the last of these Islands.* 2 vols. London.

Slocum, M. C., and F. L. Gerdel. 1965. *Vocabulario Tzeltal de Bachajon.* México: Serie de vocabularios indígenas Mariano Silva y Aceves, 13.

Smith, N. J. H. 1976. "Utilization of Game along Brazil's Transamazon Highway." *Acta Amazonica* 6: 455–466.

Smole, W. 1976. *The Yanoama Indians: A Cultural Geography.* Austin.

Soares de Souza, Gabriel. 1945. *Notícia do Brasil* [1587]. Edited by Piraja da Silva, 2 vols. São Paulo.

Solari, B. T. 1928. *Ensayo de filología: Breve vocabulario Español–Guaraní, con las relaciones etimológicas del idioma americano.* Buenos Aires.

Soukup, J. 1961. "Materiales para el catálogo de los mamíferos peruanos." *Biota* 3: 31–44, 68–84, 133–161, 240–276, 277–324, 325–331.

Southey, Robert. 1810–1819. *History of Brazil.* 3 vols. London.

Sowls, L. K. 1961. "Gestation Period of the Collared Peccary." *Journal of Mammalogy* 42: 425–426.

—— 1966. "Reproduction in the Collared Peccary (*Tayassu tajacu*)." In: I. W. Rowlands, ed. *Comparative Biology of Reproduction in Mammals* (London), pp. 155–172.

—— 1969. "The Collared Peccary." *Animals* 12, 5: 218–223.

—— 1974. "Social Behaviour of the Collared Peccary, *Dicotyles tajacu* L." In: V. Geist and F. Walther, eds. *The Behaviour of Ungulates and its Relation to Management.* Morges, Switzerland: International Union for the Conservation of Nature and Natural Resources, Publication n. s. 24, pp. 144–165.

Spicer, E. H. 1980. *The Yaquis: A Cultural History.* Tucson.

Spores, R. 1965. "The Zapotec and Mixtec at Spanish Contact." *Handbook of Middle American Indians.* Edited by G. R. Willey. 3: pp. 962–987. Austin and London.

Squier, E. G. 1969. *Notes on Central America, particularly the States of Honduras and Salvador* [1855]. New York.

Staden, Johann von. 1557. *Wahrhaftig Historia.* Marpurg.

—— 1592. *Historiae Brasilianae Ionnis Stadii Hessi.* In: Theodor de Bry *Americae Tertia Pars.* Francofurti ad Moenum.

—— 1874. *The Captivity of Hans Stade of Hesse in A.D. 1547–1555 among the Wild Tribes of Eastern Brazil.* Edited by R. F. Burton. Translated by A. Tootal. London: Hakluyt Society, 51.

—— 1963. *Zwei Reisen nach Brasilien, 1548–1555.* Edited by K. Fouquet. Marburg am der Lahn.

Starr, F. 1902. "Notes upon the Ethnography of Southern Mexico: Expedition of 1901." *Proceedings of the Davenport Academy of Sciences* 9: 1–109.

Stedman, J. G. 1796. *Narrative of a Five Years' Expedition against the Revolted Negroes of Surinam.* 2 vols. London.

—— 1963. *Expedition to Surinam* [1772–1777]. Edited by C. Bryant. London.

Steere, J. B. 1903. *Narrative of a Visit to Indian Tribes of the Purus River, Brazil.* Washington, D.C.: Carnegie Institution, Report of the United States National Museum for 1901, pp. 359–393.

Steggerda, M. 1941. *Maya Indians of Yucatán.* Washington, D.C.: Carnegie Institution, Publication 531.

Steinen, Karl von den. 1892. *Die Bakaïrí-Sprache.* Leipzig.

—— 1904. *Diccionario Sipibo.* Berlin.

Stempell, W. 1908. "Die Tierbilder der Mayahandschriften." *Zeitschrift für Ethnologie* 40: 704–743.

Steward, J. H. 1948a. "The Circum-Caribbean Tribes: An Introduction." In: *Handbook of South American Indians.* Edited by J. H. Steward. Washington, D.C.: Smithsonian Institution, Bureau of American Ethnology, Bulletin 143, 4: 1–41.

—— 1948b. "Western Tucanoan Tribes." In: *Handbook of South American Indians.* Edited by J. H. Steward. Washington, D.C.: Smithsonian Institution, Bureau of American Ethnology, Bulletin 143, 3: 737–748.

—— 1948c. "The Witotoan Tribes." In: *Handbook of South American Indians.* Edited by J. H. Steward. Washington, D.C.: Smithsonian Institution, Bureau of American Ethnology, Bulletin 143, 3: 749–762.

—— 1948d. "The Tribes of the Montaña: An Introduction." In: *Handbook of South American*

Indians. Edited by J. H. Steward. Washington, D.C.: Smithsonian Institution, Bureau of American Ethnology, Bulletin 143, 3: 507–533.

Steward, J. H., and A. Métraux. 1948a. "Tribes of the Peruvian and Ecuadorian Montaña." In: *Handbook of South American Indians.* Edited by J. H. Steward. Washington, D.C.: Smithsonian Institution, Bureau of American Ethnology, Bulletin 143, 3: 535–656.

—— 1948b. "The Peban Tribes." In: *Handbook of South American Indians.* Edited by J. H. Steward. Washington, D.C.: Smithsonian Institution, Bureau of American Ethnology, Bulletin 143, 3: 727–736.

Stirling, M. W. 1938. *Historical and Ethnographical Material on the Jívaro Indians.* Washington, D.C.: Smithsonian Institution, Bureau of American Ethnology, Bulletin 117.

Stoll, O. 1884. *Zur Ethnographie der Republik Guatemala.* Zurich.

—— 1887. *Die Sprache der Ixil-Indianer.* Leipzig.

—— 1888. *Die Maya-Sprachen der Pokom-Gruppe* (Wien) 1 [Pokonchí-Indianer].

—— 1889. *Die Ethnologie der Indianerstämme von Guatemala.* Supplement to Internationales Archiv für Ethnographie 1. Leiden.

Stone, Doris Z. 1949. *The Boruca of Costa Rica.* Papers of the Peabody Museum of American Archaeology and Ethnology, Harvard University, 26, 2.

—— 1962. *The Talamancan Tribes of Costa Rica.* Papers of the Peabody Museum of American Archaeology and Ethnology, Harvard University, 43, 2.

—— 1966. "Synthesis of Lower Central American Ethnohistory." *Handbook of Middle American Indians.* Edited by G. F. Ekholm and G. R. Willey. Austin and London. 4: 209–233.

Stout, D. B. 1947. *San Blas Cuna Acculturation: An Introduction.* Viking Fund Publications in Anthropology, 9. New York.

—— 1948a. "The Cuna." In: *Handbook of South American Indians.* Edited by J. H. Steward. Washington, D.C.: Smithsonian Institution, Bureau of American Ethnology, Bulletin 143, 4: 257–268.

—— 1948b. "The Chocó." In: *Handbook of South American Indians.* Edited by J. H. Steward. Washington, D.C.: Smithsonian Institution, Bureau of American Ethnology, Bulletin 143, 4: 269–276.

Stradelli, E. 1928. "Vocabulários da Lingua Geral, Portuguez–Nheêngatú e Nheêngatú–Portuguez." *Revista do Instituto Histórico e Geográfico Brasileiro* 158, 2: 5–768.

Stresser-Péan, G. 1952–1953. "Les Indiens Huasteques." In: I. Bernal and E. D. Hurtado, eds. *Huastecos, Totonacos y sus Vecinos.* Sociedad Méxicana de Antropología, México. Pp. 213–234.

Strömer, C. von. 1932. *Die Sprache der Mundurukú.* Wien: Anthropos: Collection Internationale de Monographies Linguistiques, 11.

Stuart, L. C. 1964. "Fauna of Middle America." *Handbook of Middle American Indians.* Edited by R. C. West. Austin and London. 1: 316–362.

Tapia Zenteno, Carlos de. 1767. *Noticia de la lengua Huastica.* México.

Tastevin, C. 1923. "Nomes de plantas e animaes em lingua Tupi." *Revista do Museu Paulista* 13: 687–764.

—— 1926. "Le Haut Tarauacá." *La Géographie* 45: 34–54, 158–175.

Tate, G. H. H. 1939. "The Mammals of the Guiana Region." *Bulletin of the American Museum of Natural History* 76: 151–229.

Tauste, Francisco de. 1888. *Arte Bocabulario . . . de la lengua de Cumana* [1680]. In: J. Platzmann, ed. *Algunas obras raras sobre la lengua Cumanagota.* 5 vols. Leipzig. 1.

Taylor, D. 1938. *The Caribs of Dominica.* Washington, D.C.: Smithsonian Institution, Bureau of American Ethnology, Bulletin 119, Anthropological Paper 3, pp. 103–160.

Taylor, K. I. 1972. "Sanumá (Yanoama) Food Prohibitions: The Multiple Classifications of Society and Fauna." Unpublished Ph.D. thesis, University of Wisconsin.

—— 1974. *Sanuma Fauna: Prohibitions and Classifications.* Caracas: Fundación La Salle de Ciencias Naturales.

—— 1981. "Knowledge and Praxis in Sanumá Food Prohibitions." In: K. M. Kensinger and W. H. Kracke, eds. *Food Taboos in Lowland South America.* Working Papers on South American Indians. Bennington College, Vermont. 3: 24–54.

Temple, R. 1860. [Letter to P. L. Sclater, Secretary of the Zoological Society of London] *Proceedings of the Zoological Society of London* 1860: 206–207.

Tertre, Jean-Baptiste du. 1667–1671. *Histoire générale des Antilles habitées par les François.* 3 vols. Paris.

Thayer, J. E., and O. Bangs. 1905. "The Mammals and Birds of the Pearl Islands, Bay of Panama." *Bulletin of the Museum of Comparative Zoology* 46, 8: 135–160.

Thévet, André. 1558. *Les singularitez de la France Antarctique*. Paris.

—— 1568. *The New Found Worlde or Antarctike*. Translated from the French. London.

—— 1575. *La Cosmographie Universelle*. 2 vols. Paris.

Thomas [Le Sieur]. 1674. *Description de l'isle de la Jamaique*. In: H. Justellus, ed. *Recueil de divers voyages faits en Afrique et en l'Amerique*. Paris. 27 pp.

Thomas, O. 1902. "On Some Mammals from Coiba Island, off the West Coast of Panama." *Novitates Zoologicae* 9: 135–137.

—— 1903a. "On a Collection of Mammals from the Small Islands off the Coast of Western Panama." *Novitates Zoologicae* 10: 34–42.

—— 1903b. "On the Mammals collected by Mr. A. Robert at Chapada, Mato Grosso (Percy Sladen Expedition to Central Brazil)." *Proceedings of the Zoological Society of London* 1903: 232–244.

—— 1927. "On Mammals from the Upper Huallaga and Neighbouring Highlands." *Annals and Magazine of Natural History* 20: 594–608.

Thompson, J. E. S. 1970. *Maya History and Religion*. Norman.

—— 1972. *A Commentary on the Dresden Codex: A Maya Hieroglyphic Book*. Memoirs of the American Philosophical Society 93. Philadelphia.

Thord-Gray, I. 1955. *Tarahumara-English, English-Tarahumara Dictionary*. Coral Gables, Florida.

Thurn, E. F. Im. 1883. *Among the Indians of Guiana*. London.

Tomes, R. F. 1861. "Report of a Collection of Mammals at Duenas, Guatemala." *Proceedings of the Zoological Society of London* 1861: 278–288.

Topsell, Edward. 1607. *The Historie of Foure-footed Beastes, collected out of all the Volumes of Conradus Gesner, and all other Writers to this Present Day*. London.

Toribio de Ortiguera. 1968. *Jornada del Río Marañon* [ca. 1580–1590]. Edited by Mario Hernández Sánchez-Barra. Madrid: Biblioteca de Autores Españoles, 216.

Torquemada, Juan de. 1723. *Monarquia Indiana* [1615]. 3 vols. Madrid.

Torres de Araúz, R. 1977. "Las culturas indígenas Panameñas en el momento de la conquista." *Hombre y Cultura* 3: 69–96.

Tozzer, A. M. 1907. *A Comparative Study of the Mayas and the Lacandones*. New York: Archaeological Institute of America.

Tozzer, A. M., and G. M. Allen. 1910. *Animal Figures in the Maya Codices*. Papers of the Peabody Museum of American Archaeology and Ethnology, Harvard University, 4, 3: 275–372.

Treutlein, T. E. 1945. "The Relation of Felipp Ségesser [1737]." *Mid-America* 27 [n. s. 16]: 137–187, 257–260.

Trimborn, H. 1952. "Pascual de Andagoya on the Cueva of Panama." In: S. Tax, ed. *Indian Tribes of Aboriginal America: Selected Papers of the 29th International Congress of Americanists*. Chicago. Pp. 254–261.

Trouessart, E. L. 1897–1899. *Catalogus Mammalium*. 2 vols. Berolini.

—— 1904. *Catalogus Mammalium: Quinquennale Supplementum*. Berolini.

Tschudi, J. J. von. 1844–1846. *Untersuchungen über die Fauna Peruana*. 2 vols. St. Gallen.

—— 1847. *Travels in Peru during the years 1838–1842*. Translated from the German. London.

—— 1853. *Die Kechua-Sprache*. 3 vols. Wien.

Turner, H. N. 1849. "On the Evidences of Affinity afforded by the Skull in the Ungulate Mammalia." *Proceedings of the Zoological Society of London* 1849: 147–158.

Turner, Joan S. 1967. "Environment and Cultural Classification: A Study of the Northern Kayapó." Unpublished Ph.D. thesis, Harvard University.

Tyson, Edward. 1683. "Tajacu seu Aper Mexicanus Moschiferus." *Philosophical Transactions of the Royal Society of London* 13, 153: 359–382.

Up de Graff, F. 1974. "Jívaro Field Clearing with Stone Axes." In: P. J. Lyon, ed. *Native South Americans*. Boston. Pp. 120–122.

Usher, G. 1974. *Dictionary of Plants used by Man*. London.

Valmont de Bomare, J. C. 1767–1768. *Dictionnaire raisonné universel d'histoire naturelle*. 6 vols. Paris.

Vanzolini, P. E. 1956–1958. "Notas sôbre a zoologia dos indios Canela." *Revista do Museu Paulista* n. s. 10: 155–171.

Vargas Machuca, Bernardo de. 1599. *Milicia y descripción de las Yndias*. Madrid.

Vázquez de Espinosa, Antonio. 1942. *Compendio y descripción de las Indias Occidentales.* Edited and translated [English] by C. U. Clark. Washington, D.C.: Smithsonian Institution, Miscellaneous Collections, 102, [text in Spanish, ibid., 108, 1948].

Veigl, Franz Xavier. 1785. *Gründliche Nachrichten über die Verfassung der Landschaft von Maynas in Sud-Amerika bis zum Jahre 1768.* In: C. G. von Murr, ed. *Reisen einiger Missionarien der Gesellschaft Jesu in Amerika.* Nürnberg. Pp. 1–450.

Velasco, Juan de. 1946. *Historia del Reino de Quito* [1789]. 3 vols. Quito.

Vellard, J. 1939. *Une civilisation du miel: Les indiens Guayakis du Paraguay.* Paris.

Vespucci, Amerigo. 1893. *The First Four Voyages of Amerigo Vespucci.* Facsimile of 1st edition [1505–1506, Florence] and translation, London.

—— 1894. *The Letters of Amerigo Vespucci and Other Documents illustrative of His Career.* Edited by C. R. Markham. London: Hakluyt Society, 90.

Vetancurt, Agustín de. 1960–1961. *Teatro Mexicano* [1698]. 4 vols. México.

Vickers, W. T. 1976. "Cultural Adaptation to Amazonian Habitats: The Siona-Secoya of Eastern Ecuador." Unpublished Ph.D. thesis, University of Florida.

—— 1979. "Native Amazonian Subsistence in Diverse Habitats: The Siona-Secoya of Ecuador." In: E. F. Moran, ed. *Changing Agricultural Systems in Latin America.* Studies in Third World Societies, 7: 6–36. Williamsburg.

—— 1980. "An Analysis of Amazonian Hunting Yields as a Function of Settlement Age." In: R. B. Hames, ed. *Studies in Hunting and Fishing in the Neotropics.* Working Papers on South American Indians. Bennington College, Vermont. 2: 7–29.

Villa, Bernardo. 1948. "Maníferos del Soconusco, Chiapas." *Anales del Instituto de Biología* [México] 19, 2: 485–528.

Villa Rojas, A. 1945. *The Maya of East Central Quintana Roo.* Washington, D.C.: Carnegie Institution, Publication 559.

—— 1969a. "Maya Lowlands: The Chontal, Chol, and Kekchi." *Handbook of Middle American Indians.* Edited by E. Z. Vogt. Austin and London. 7, 1: 230–243.

—— 1969b. "The Tzeltal." *Handbook of Middle American Indians.* Edited by E. Z. Vogt. Austin and London. 7, 1: 195–225.

—— 1969c. "The Maya of Yucatan." *Handbook of Middle American Indians.* Edited by E. Z. Vogt. Austin and London. 7, 1: 244–275.

Villas Boas, O., and C. Villas Boas. 1974. *Xingu: The Indians, Their Myths.* London.

Vogt, E. Z. 1970. *Zinacantecos of Mexico.* New York.

Vogt, F. 1904. "Die Indianer des Obern Paraná." *Mitteilungen der Anthropologischen Gesellschaft in Wien* 34 [ser. 3, 4]: 200–221.

Von Hagen, V. W. 1939. *The Tsátchela Indians of Western Ecuador.* New York: Museum of the American Indian, Heye Foundation, Indian Notes and Monographs 51.

—— 1943. *The Jicaque (Torrupan) Indians of Honduras.* New York: Museum of the American Indian, Heye Foundation, Indian Notes and Monographs 53.

Vos, A. de., R. H. Manville, and R. G. Van Gelder. 1956. "Introduced Mammals and their Influence on Native Biota." *Zoologica* 41: 163–194.

W. [M.] 1732. *The Mosquito Indian and His Golden River, being A Familiar Description of the Mosqueto Kingdom in America.* In: A. Churchill, ed. *A Collection of Voyages and Travels.* 8 vols., 1704–1752. London. 6: 285–298.

Wafer, Lionel. 1934. *A New Voyage and Description of the Isthmus of America* [1699]. Edited by L. E. Elliott Joyce. London: Hakluyt Society, 2d ser., 73.

Wagley, C. 1977. *Welcome of Tears: The Tapirapé Indians of Central Brazil.* Oxford.

Wagley, C., and E. Galvão. 1948a. "The Tenetehara." In: *Handbook of South American Indians.* Edited by J. H. Steward. Washington, D.C.: Smithsonian Institution, Bureau of American Ethnology, Bulletin 143, 3: 137–148.

—— 1948b. "The Tapirapé." In: *Handbook of South American Indians.* Edited by J. H. Steward. Washington, D.C.: Smithsonian Institution, Bureau of American Ethnology, Bulletin 143, 3: 167–178.

—— 1949. *The Tenetehara Indians of Brazil: A Culture in Transition.* Columbia University Contributions to Anthropology 35. New York.

Wagner, P. L. 1958. *Nicoya: A Cultural Geography.* Berkeley and Los Angeles: University of California Publications in Geography 12, 3.

Walker, E. P. 1975. *Mammals of the World.* 2 vols., 3d ed. Baltimore.

Wallace, A. R. 1853. *A Narrative of Travels on the Amazon and Rio Negro, with An Account of the Native Tribes and Observations on the Climate, Geology and Natural History of the Amazon Valley.* London.

—— 1876. *The Geographical Distribution of Animals*. 2 vols. London.

Warren, George. 1752. *An Impartial Description of Surinam upon the Continent of Guiana in America*. In: A. Churchill, ed. *A Collection of Travels and Voyages*. 8 vols., 1704–1752. London. 8: 920–931.

Wassén, H. 1935. "Notes on Southern Groups of Chocó Indians of Colombia." *Etnologiska Studier* 1: 35–182.

Wavrin, Robert [Marquis de]. 1932. "Folklore du Haut-Amazone." *Journal de la Société des Américanistes de Paris* n. s. 24: 121–146.

—— 1939. *Les Bêtes sauvages de l'Amérique*. Paris.

Weaver, Murial P. 1972. *The Aztecs, Maya and their Predecessors*. New York and London.

Wegner, R. N. 1936. *Zum sonnentor durch altes Indianerland*. Darmstadt.

Weiss, G. 1972. "Campa Cosmology." *XXXIX Congreso Internacional de Americanistas* (Lima, 1970), *Actas y Memorias* (Lima) 4: 189–206.

Wernicke, E. 1938. "Rutas y etapas de las introducción de los animales domésticos en las tierras Americanas." *Gaea: Anales de la Sociedad Argentina de Estudios Geográficos* 1938: 77–83.

West, R. C. 1957. *The Pacific Lowlands of Colombia: A Negroid Area of the American Tropics*. Baton Rouge: Louisiana State University Studies, Social Science Series 8.

—— 1964. "The Natural Regions of Middle America." *Handbook of Middle American Indians*. Edited by R. C. West. Austin and London. 1: 363–383.

Wetzel, R. M. 1977a. "The Chacoan Peccary, *Catagonus wagneri* (Rusconi)." *Bulletin of the Carnegie Museum of Natural History* 3: 1–36.

—— 1977b. "The Extinction of Peccaries and a New Case of Survival." *Annals of the New York Academy of Sciences* 288: 536–544.

Wetzel, R. M., R. E. Dubos, R. L. Martin, and P. Myers. 1975. "*Catagonus* an 'extinct' Peccary, alive in Paraguay." *Science* 189: 379–381.

Whiffen, T. 1913. "A Short Account of the Indians of the Issá-Yapurá District (South America)." *Folklore* 1913: 41–62.

—— 1915. *The North-West Amazons*. London.

White, T. E. 1952–1953. "A Method of Calculating the Dietary Percentage of Various Food Animals Utilized by Aboriginal Peoples." *American Antiquity* 18: 396–398.

Whiton, L. C., H. B. Greene, and R. P. Momsen. 1964. "The Isconahua of the Remo." *Journal de la Société des Américanistes de Paris* 53: 85–124.

Whitten, N. E. 1976. *Sacha Runa: Ethnicity and Adaptation of Ecuadorian Jungle Quichua*. Urbana and London.

Wilbert, J. 1970. *Folk Literature of the Warao Indians: Narrative Material and Motif Content*. Los Angeles: Latin American Center, University of California.

—— 1972. *Survivors of El Dorado: Four Indian Cultures of South America*. New York.

—— 1974. *Yupa Folktales*. Los Angeles: Latin American Center, University of California.

—— 1978. *Folk Literature of the Gé Indians*. Los Angeles: Latin American Center, University of California, 1.

Wilbert, J., and K. Simoneau. 1982. *Folk Literature of the Mataco Indians*. Los Angeles: Latin American Center, University of California.

Willey, G. R. 1978. "Artifacts." In: G. R. Willey, ed. *Excavations at Seibal*. Memoirs of the Peabody Museum of American Archaeology and Ethnology, Harvard University, 14, 1.

Willey, G. R., W. R. Bullard, J. B. Glass, and J. C. Gifford. 1965. *Prehistoric Maya Settlements in the Belize Valley*. Papers of the Peabody Museum of American Archaeology and Ethnology, Harvard University, 54.

Willey, G. R., and C. R. McGimsey. 1954. *The Monagrillo Culture of Panama*. Papers of the Peabody Museum of American Archaeology and Ethnology, Harvard University, 49, 2.

Williams, J. 1932. *Grammar, Notes and Vocabulary of the Language of the Makuchi Indians of Guiana*. St. Gabriel-Mödling (Austria): Anthropos: Collection Internationale de Monographies Linguistiques.

Williams, J. J. 1852. *The Isthmus of Tehuantepec*. New York.

Wing, E. S. 1975. "Animal Remains from Lubaantun." In: N. Hammond, *Lubaantun: A Classic Maya Realm*. Harvard University, Peabody Museum Monograph 2: 379–383.

—— 1977. "Factors influencing Exploitation of Marine Resources." In: E. P. Benson, ed. *The Sea in the Pre-Columbian World*. Washington, D.C.: Dumbarton Oaks, 47–66.

Wing. E. S., and D. Steadman. 1980. "Vertebrate Faunal Remains from Dzibilchaltun, Yucatan." In: E. W. Andrews IV and E. W. Andrews V, eds. *Excavations at Dzibilchaltun, Yucatan*, Middle American Research Institute, Publication 48: 326–331.

Wissmann, H. W. L. L. 1891. *Meine zweite Durchquerung Aequatorial Afrikas.* Frankfurt am Oder.

Woodburne, M. O. 1968. "The Cranial Myology and Osteology of *Dicotyles tajacu*, the Collared Peccary, and its bearing on Classification." *Memoirs of the Southern California Academy of Science* 7: 1–48.

—— 1969. "A Late Pleistocene Occurrence of the Collared Peccary, *Dicotyles tajacu*, in Guatemala." *Journal of Mammalogy* 50, 1: 121–125.

Woodroffe, J. F. 1914. *The Upper Reaches of the Amazon.* London.

Ximénez, Francisco. 1967. *Historia natural del reino de Guatemala* [1722]. Guatemala.

Yangues, Manuel de. 1888. *Principios y reglas de la lengua Cumanagota . . . con un diccionario* [1683]. In: J. Platzmann, ed. *Algunas obras raras sobre la lengua Cumanagota.* 5 vols. Leipzig. 2.

Yde, J. 1948. "The Regional Distribution of South American Blowgun Types." *Journal de la Société des Américanistes de Paris* n. s. 37: 275–317.

—— 1965. *Material Culture of the Waiwái.* Copenhagen.

Youatt, W. 1860. *The Pig.* London.

Young, P. D. 1971. *Ngawbe: Tradition and Change among the Western Guaymí of Panama.* Urbana: University of Illinois Studies in Anthropology 7.

Young, Thomas. 1842. *Narrative of a Residence on the Mosquito Shore during the years 1839 to 1841.* London.

Zerries, O. 1954. *Wild-und Buschgeister in Südamerika: Eine Untersuchung jägerzeitlicher Phänomene im Urbild südamerikanischer Indianer.* Wiesbaden: Studien zur Kulturkunde 11.

—— 1968. "Primitive South America and the West Indies." In: W. Krickeberg, H. Trimborn, W. Müller, and O. Zerries, *Pre-Columbian Religions.* Translated by S. Davis. London. Pp. 230–316.

Zevallos, Agustín de. 1886. *Memorial para el Rey Nuestro Señor de la descripción y calidades de la provincia de Costa Rica, año de 1610.* In: D. León Fernández, ed. *Colección de documentos para la historia de Costa Rica.* 10 vols., 1881–1907. San José, Paris, and Barcelona. 5: 156–161.

Zimmermann, E. A. W. 1777. *Specimen Zoologiae Geographicae Quadrupedum.* Lugduni Batavorum.

Zuckerman, S. 1953. "The Breeding Seasons of Mammals in Captivity." *Proceedings of the Zoological Society of London* 122: 827–950.

SUPPLEMENT TO BIBLIOGRAPHY

The present monograph was submitted for publication in July 1983. It has not been possible to take account of two important recent publications: H. B. Hames and W. T. Vickers (eds), *Adaptive Responses of Native Amazonians*, New York and London, 1983, and L. K. Sowls, *The Peccaries*, Tucson, 1984.

I would also like to draw attention to: T. de Booy, *Notes on the Archaeology of Margarite Island, Venezuela*, Contributions from the Museum of the American Indian, Heye Foundation, II (5), New York, 1916 (p. 16, bones of peccary); M. R. Harrington, *Cuba before Columbus*, Indian Notes and Monographs, Miscellaneous Series 17, Museum of the American Indian, Heye Foundation, New York, 2 vols., 1921 (1: p. 164 "a bone of a peccary . . . . in an aboriginal deposit"); P. Kelemen, *Medieval American Art*, 2 vols., New York, 1944–1946 (1: p. 348, 2: pl. 284, incised peccary skull from Copan, Honduras, and illustration of a group of running peccaries); and E. Magaña, "Carib Myths about the Origin of some Animal Species," *Latin American Indian Literatures Journal* 1 (1), 1985: pp. 13–27 (pp. 13–14, 16–19, human beings transformed into white-lipped peccaries).

INDEX

PUBLICATIONS

OF

The American Philosophical Society

The publications of the American Philosophical Society consist of PROCEEDINGS, TRANS-ACTIONS, MEMOIRS, and YEAR BOOK.

THE PROCEEDINGS contains papers which have been read before the Society in addition to other papers which have been accepted for publication by the Committee on Publications. In accordance with the present policy one volume is issued each year, consisting of four numbers, and the price is $20.00 net per volume. Individual copies of the PROCEEDINGS are $10.00.

THE TRANSACTIONS, the oldest scholarly journal in America, was started in 1769. In accordance with the present policy each annual volume is a collection of monographs, each issued as a part. The current annual subscription price is $60.00 net per volume. Individual copies of the TRANSACTIONS are offered for sale.

Each volume of the MEMOIRS is published as a book. The titles cover the various fields of learning; most of the recent volumes have been historical. The price of each volume is determined by its size and character, but subscribers are offered a 20 per cent discount.

The YEAR BOOK is of considerable interest to scholars because of the reports on grants for research and to libraries for this reason and because of the section dealing with the acquisitions of the Library. In addition it contains the Charter and Laws, and lists of members, and reports of committees and meetings. The YEAR BOOK is published about April 1 for the preceding calendar year. The current price is $8.50.

An author desiring to submit a manuscript for publication should send it to the Editor, American Philosophical Society, 104 South Fifth Street, Philadelphia, Pa. 19106.

www.ingramcontent.com/pod-product-compliance
Lightning Source LLC
Chambersburg PA
CBHW061755260326
41914CB00006B/1119